TRANSFORMING URBAN–RURAL WATER LINKAGES INTO HIGH-QUALITY INVESTMENTS

AUGUST 2020

ADB

ASIAN DEVELOPMENT BANK

© 2020 Asian Development Bank
6 ADB Avenue, Mandaluyong City, 1550 Metro Manila, Philippines
Tel +63 2 8632 4444; Fax +63 2 8636 2444
www.adb.org

Some rights reserved. Published in 2020.

ISBN 978-92-9262-331-9 (print), 978-92-9262-332-6 (electronic), 978-92-9262-333-3 (ebook)
Publication Stock No. TCS200228-2
DOI: http://dx.doi.org/10.22617/TCS200228-2

The views expressed in this publication are those of the authors and do not necessarily reflect the views and policies of the Asian Development Bank (ADB) or its Board of Governors or the governments they represent.

ADB does not guarantee the accuracy of the data included in this publication and accepts no responsibility for any consequence of their use. The mention of specific companies or products of manufacturers does not imply that they are endorsed or recommended by ADB in preference to others of a similar nature that are not mentioned.

By making any designation of or reference to a particular territory or geographic area, or by using the term "country" in this document, ADB does not intend to make any judgments as to the legal or other status of any territory or area.

Please contact pubsmarketing@adb.org if you have questions or comments with respect to content, or if you wish to obtain copyright permission for your intended use that does not fall within these terms, or for permission to use the ADB logo.

Corrigenda to ADB publications may be found at http://www.adb.org/publications/corrigenda.

Notes:
In this publication, "$" refers to United States dollars.
ADB recognizes "China" as the People's Republic of China.

On the cover: The schematic diagram presents the important elements of urban-rural water linkages (cover graphics by Claudette Rodrigo).

Contents

Tables, Figures, Boxes, and Maps iv
Foreword v
Acknowledgments vii
Abbreviations viii
Weights and Measures viii
Executive Summary ix

I.   Improving River Health by Integrated Urban–Rural Management   1
    A.   Background   1
    B.   Global and Regional Water Challenges   2
    C.   Water Resources Issues in the People's Republic of China   4
    D.   Urban–Rural Linkages and the Need For Integration   9

II.   Urban–Rural Water Linkages in Tuan River, Dengzhou City   14
    A.   Dengzhou City   14
    B.   Tuan River Basin   14
    C.   Land Use Impacts   14
    D.   Water Withdrawal and Water Use   19
    E.   Pollution   20
    F.   Taking Action: Addressing the Urban–Rural Water Issues in Dengzhou City   29

III.   Government Strategies, Policies, and Initiatives   30
    A.   Socioeconomic Urban–Rural Challenges in the People's Republic of China   30
    B.   Water Resources Management Challenges and Issues in the People's Republic of China   32
    C.   Water and Natural Resources Management: Laws, Policies, and Strategies   32
    D.   Yangtze River Economic Belt Development Program   34
    E.   Yellow River Basin Program   35
    F.   Institutional Reform   35
    G.   Assessing Urban–Rural Water Linkages in Government Programs   35

IV.   Embedding Urban–Rural Water Linkages in Project Design   37
    A.   Land Use   38
    B.   Surface Water and Groundwater Withdrawal   42
    C.   Pollution   43
    D.   Institutional Strengthening   44
    E.   Assessing Project Impacts   46

V.   Conclusion and Way Forward   51
    A.   Summary of Urban–Rural Water Linkages for Dengzhou City   51
    B.   Features of the Project Interventions and Key Values Added   51
    C.   Way Forward and Upscaling Experience Gained   53

Tables, Figures, Boxes, and Maps

Tables

1	Land Use in the Tuan River Basin, 2015	23
2	Pollution Reduction by Key Project Components	49
3	Pollutant Interception by Bio-shield Buffer Zone, Tuan River	50
4	Summary of Pollutant Reduction by the Project	50

Figures

1	Some Key Water Resources Issues in the People's Republic of China	6
2	Projections of National Water Supply and Demand	7
3	River Basin Flow Changes, 1980–2014	8
4	Water Quality Trend in the Main Rivers of the People's Republic of China, 2003–2014	8
5	Schematic of Identified Urban–Rural Water Linkages, Tuan River Basin	11
6	Categories of Influences of the Urban–Rural Water Linkages	12
7	Pollutant Analysis of the Tuan River	21
8	Illustration of Point and Nonpoint Sources of Pollution	21
9	Impact of Pollutants on the Food Supply Chain	28
10	Layout of the Proposed Tuanbei New District Green Corridor Park	39
11	Internal Landscape Design of the Proposed Tuanbei New District Green Corridor Park	39
12	Design of Tuan River Improvement Works	41
13	Dengzhou City Government's Asset Management System for Water Infrastructure	48

Boxes

1	Water Security in the Asian Region	3
2	Water Security in the Yellow River Basin	7
3	Key Features of the Proposed Asset Management System	47
4	Potential Application of the Urban–Rural Water Linkages Approach in the Pacific Islands	55

Maps

1	Project Location—Dengzhou City in Nanyang City of Henan Province	15
2	Aerial View of Dengzhou City	16
3	Locations of Key Point and Nonpoint Sources of Pollution in the Tuan River Basin	22
4	Linkages between Waste Pollution and River Recharge in the Tuan River Downstream of Dengzhou	26

Foreword

At the time of writing this foreword, the global health crisis stemming from the coronavirus disease (COVID-19) pandemic is causing unprecedented human suffering and economic losses, affecting disproportionally the most vulnerable. The lessons learnt from this pandemic will hopefully reshape the relationship of humans with the natural environment to be more symbiotic or harmonious than in the past.

Nowhere has recent economic growth been more rapid than in the People's Republic of China (PRC). Significant improvements in human well-being and quality of life have taken place at the expense of the health of rivers, lakes, and fisheries. Many rivers across the PRC are highly degraded, transporting high loads of pollutants and nutrients from industries, agriculture, and urban centers. Some 30% of water resources along the main rivers in the PRC is unsuitable for human use due to poor quality. Climate change pressures on drying river flows cause additional environmental health impacts and increase the likelihood of transmission of waterborne diseases due to poor sanitation and wastewater management.

The PRC has anchored its path toward ecological protection and high-quality development to its vision of *ecological civilization*, a concept first introduced in 2007 and increasingly emphasized in the country's subsequent 5-year plans. The growing awareness and political commitment to conserving and protecting the nation's water resources are reflected in recent institutional reforms (including the "three red lines" policy and the "ecological redline" policy, which focus on strengthening regulatory frameworks), along with implementation mechanisms to protect ecosystem services. By also recognizing the need to tackle climate change, these reforms and measures provide the basis for more innovative, holistic, and sustainable approaches to development under the 14th Five-Year Plan, 2021–2025.

This report sheds some light on the PRC's approach to restoring the eco-environmental conditions of the seriously degraded Tuan River in Henan Province. Referred to as the urban–rural water linkages (URWLs) approach, the Asian Development Bank (ADB) designed four clusters of interventions to simultaneously address (i) inappropriate land use, (ii) depletion of river flows and groundwater, (iii) sources of pollution, and (iv) institutional constraints. ADB and the Government of the PRC both recognize the direct connection between urban and rural areas through rivers and groundwater aquifers, together with the interdependency of urban and rural land use, water use, and sources of pollution. The URWLs approach also has the advantage of improving services and opportunities, consistent with the PRC's rural vitalization strategy.

The Tuan River is a second-order tributary of the Han River, which is a major tributary of the Yangtze River. The UWRLs approach is implemented through the Henan Dengzhou Integrated River Restoration and Ecological Protection Project. The project is part of the Yangtze River Economic Belt (YREB) program, which is the ADB–PRC centerpiece of cooperation to enable socioeconomic growth and environmental protection in the upper and middle Yangtze River Basin. Building upon the success of the YREB program, the Government of the PRC, in partnership with ADB, is designing a similar program for the Yellow River Basin, which will advance ecological protection and the URWLs approach in the upcoming 14th Five-Year Plan.

Drawing upon the PRC's experience and lessons learnt, this publication seeks to share the URWLs approach with ADB's developing member countries across Asia and the Pacific. By doing so, countries can adopt more sustainable solutions for remediating and rehabilitating rivers, as well as pursue a more balanced relationship between humans and the natural environment.

James Lynch
Director General
East Asia Department
Asian Development Bank
Manila, Philippines

Acknowledgments

The author, Rabindra P. Osti, senior water resources specialist from the East Asia Department (EARD) of the Asian Development Bank (ADB), would like to thank Qingfeng Zhang, director of EARD's Environment, Natural Resources, and Agriculture Division (EAER), for his supervision and comments during the preparation of this report; M. Teresa Kho, deputy director general, EARD, for providing guidance and discerning advice; and James Lynch, director general, EARD, for supporting and endorsing this publication.

From the Office of the Director General, EARD, Akiko Terada-Hagiwara, principal economist, and Yumiko Tamura, principal operations coordination specialist, provided overall knowledge management support; while Sophia Castillo-Plaza organized the briefing seminar attended by ADB staff, which presented the findings and recommendations of the draft publication.

ADB peer reviewers—Nargiza Parkhatovna Talipova, principal portfolio management specialist, Southeast Asia Department; and Jelle Beekma, senior water resources specialist, Water Sector Group Secretariat under the Sustainable Development and Climate Change Department—offered comprehensive comments and constructive recommendations at different stages of the report preparation. Xueliang Cai, water resources specialist, EAER; Silvia Cardascia, young professional, EAER; and Xiaoyan Yang, senior programs officer, People's Republic of China Resident Mission, also provided helpful insights and suggestions. The author is also indebted to John William Porter and Eelco Van Beek, senior water resources management specialists (resource persons), for their review, constructive advice, and effort in the preparation of this report.

Joy Quitazol-Gonzalez reviewed the final draft and facilitated the production process from editing, proofreading, design, and through to final publication. The author also acknowledges the valuable assistance of Margaret Clare Anosan, Noreen Joy Ruanes, and Mary Dianne Rose Umayan, EAER.

Abbreviations

ADB	–	Asian Development Bank
BOD	–	biochemical oxygen demand
COVID-19	–	coronavirus disease
DCG	–	Dengzhou City Government
IWRM	–	integrated water resources management
LID	–	low-impact development
NH_3-N	–	ammonia nitrogen
NPS	–	nonpoint source
PRC	–	People's Republic of China
RCP	–	representative concentration pathway
SDG	–	Sustainable Development Goal
SNWDP	–	South-to-North Water Diversion Project
SWM	–	solid waste management
URWL	–	urban–rural water linkage
WSP	–	water supply plant
WWTP	–	wastewater treatment plant
YREB	–	Yangtze River Economic Belt

Weights and Measures

bcm	–	billion cubic meters
ha	–	hectare
kg	–	kilogram
km	–	kilometer
km^2	–	square kilometer
L	–	liter
m^3	–	cubic meter
mcm	–	million cubic meters
mg	–	milligram

Executive Summary

The coronavirus disease (COVID-19) pandemic emerged to suddenly threaten the health and well-being of multitudes. Many lives have been lost all over the world, and economic hardship will continue for many. The balance between human society and the natural environment, routinely taken for granted, has been exposed as fragile. The scale of the social disruption and economic damages being wrought had been scarcely conceivable in January 2020 or even in early February 2020. What the post-pandemic world will be like is as yet uncertain, except that it will certainly not be the same as before. One thing that the world should learn from this catastrophe as it strives to recover is that nature must be respected. With regard to the water sector, sound management of natural resources is important, and there needs to be a sufficient harmony between human and nature—in a word, sustainability. This report, which is directed toward restoring river health in degraded environments, is highly relevant to the objective of redressing the balance between human society and the natural environment.

In river basins throughout the world, rivers connect and pass through urban and rural districts; and groundwater aquifers, which underlie urban and rural areas, are connected to the rivers. Sources of pollution occur in both urban and rural areas: point sources, such as sewer outfalls and industrial discharges, are the main sources in urban areas; while nonpoint sources are generally worse in rural areas, including runoff from agricultural lands and discharges from intensive animal husbandry. Regardless of sources, pollution adversely affects urban and rural residents, contaminating groundwater and degrading water quality of rivers and lakes that are the sources of urban and rural water supply. The river health of most rivers in developing countries is deplorable, and past efforts to address the problems have been inadequate and piecemeal. Because the sources of water and the sources of water pollution extend through urban and rural parts of river basins with visible and invisible linkages, a new approach that simultaneously tackles the problems in both urban and rural areas promises to be more effective. This new approach, the urban–rural water linkages (URWLs) approach, also relates to land use, which alters runoff and infiltration, in turn modifying characteristics of floods and droughts. Land use change can trigger accelerated rates of soil erosion; leach nutrients that degrade water quality; and cause river sedimentation that modifies the morphology of rivers and streams, which alters flood risk.

Integrated water resources management (IWRM) has long been a central principle in this regard, requiring a holistic, river basin-scale approach to water resources and environmental management with spatial and functional integration of management measures, including structural and nonstructural measures. The URWLs approach advocated in this report complements IWRM by focusing on the need for integration of interventions in urban and rural parts of river basins to secure benefits sought (i) to improve the welfare and livelihoods of urban and rural communities; (ii) to maximize prosperity and equity; and (iii) to conserve natural resources, like rivers, upon which prosperity depends while enabling their sustainable exploitation. A URWLs approach recognizes the interdependency of urban and rural society, and identifies the important linkages between quantity and quality of water resources and their management in urban and rural areas.

In Asia, irrigation and water for agriculture accounts for 80% of total water use. As populations grow and expectations of improved standards of living rise, agriculture will have to produce much more food in the coming decades. Future urbanization and industrialization will add to the burgeoning demand and competition for water, increasing competition between urban and rural water use, and conflicting with environmental water requirements. The predicted effects of future climate change arising from global warming will only make these resource problems more challenging.

In developing countries, water quality is as much of a challenge as water quantity, rendering a portion of limited available water resources unfit for human use. The URWLs approach is vital in meeting both these challenges. It is also important for future physical environment and ecological sustainability in river basins, and for meeting relevant Sustainable Development Goals set by the United Nations. The outcomes of past water resources management have often fallen short because the importance of urban–rural linkages has not been adequately considered.

This report discusses the concept of URWLs and presents know-how on embedding the concept into project design, mainly through the Henan Dengzhou Integrated River Restoration and Ecological Protection Project of the Asian Development Bank (ADB). The project demonstrates a URWLs approach to address water management issues in the Tuan River Basin, a tertiary-level subbasin in the north of the Yangtze River Basin. While some features of the problems treated by interventions of this project are specific to Dengzhou City and the Tuan River, it is still a useful model that can be adapted and applied to many river basins in the People's Republic of China (PRC) and in developing countries in general.

Chapter I describes the context and background of the URWLs approach with references to water resources issues in the PRC, and it discusses the need to integrate interventions, taking into account the URWLs that exist. It identifies three key categories of urban–rural linkages: land use, use and withdrawal of water, and water pollution. Typically, there is a substantial gap between the standard of services in urban and rural communities, and yet the ramifications of inadequate services are interlinked.

Chapter II describes the prevailing conditions in Dengzhou City and the Tuan River, using the three key categories of urban–rural linkages. Most of the land use within the Dengzhou City boundaries is rural. The urban core continues to expand, with new town centers emerging around the periphery every few years. As the city expands, peri-urban areas evolve and rural areas retreat.

Chapter III outlines relevant national strategies, policies, and initiatives in the PRC. Major challenges identified are (i) rural poverty and slow economic progress; (ii) the need for greener, more environmentally sustainable approaches to support production and commerce; (iii) inferior rural governance and provision of rural infrastructure and services; and (iv) discrepancies between urban and rural areas in education and technology, and in access to social security and finance for investments. Chapter III also notes the Yangtze River Economic Belt (YREB) Development Plan, 2016–2030, which aims to provide socioeconomic stimulus and sustainable (green) development. ADB is supporting the YREB program, and the Henan Dengzhou Integrated River Restoration and Ecological Protection Project discussed in this report is an element of that support. A similar program for the Yellow River Basin is under consideration by the Government of the PRC, and ADB is ready to provide appropriate support, in which event the URWLs approach could be useful.

Chapter IV outlines the ADB-financed Henan Dengzhou Integrated River Restoration and Ecological Protection Project in detail, proposing it as a case study for the URWLs approach. Under the category of land use, project interventions include (i) Tuanbei New District ecological and water improvement on the north bank of the Tuan River, which will provide open green space, wetlands, rain gardens, and leisure facilities; (ii) Tuan River ecological restoration, which will rehabilitate a 13.8-kilometer reach of the river downstream of the main urban area; and (iii) Xingshan forest plantation for soil erosion control. Under the category of water use and withdrawal, the project is implementing two rural water supply schemes to supply treated water to 10 rural townships. Under the category of pollution control, project interventions include (i) wastewater management projects in both urban and rural areas of Dengzhou City; (ii) a solid waste management facility in a rural township, which will pilot advanced bio-remedial technology; and (iii) vegetated riparian strips (bio-shields) along both banks of the lower Tuan River for nonpoint pollution control.

For sustainability of project benefits, the project incorporates institutional capacity strengthening, which includes the integration of government provision of water supply services to both urban residents and rural consumers in the two new rural water supply schemes. Integration under a single management agency is an important innovation to better balance service standards in urban and rural areas. The project was formulated to restore river health using the URWLs approach and to demonstrate the advantages of this approach in tackling the serious river health conditions of the Tuan River in Dengzhou City.

It is concluded in chapter V that, by applying the URWLs approach, the project will contribute to redressing the environmental problems in the Tuan River Basin and rehabilitating the Tuan River, relieve some of the relative disadvantages endured by rural residents, and enhance living conditions in the rural villages and in the Tuanbei New District. The project applies IWRM and adopts the low-impact development or the sponge city design approach to urban stormwater management and other environmental innovations, and is a worthy contribution to the YREB program. Experience gained from this project will enable the URWLs approach to be refined and adapted to other river basins with varying hydrological and environmental conditions.

Improving River Health by Integrated Urban–Rural Management

A. Background

1. Many cities throughout the world are located on or near rivers—and for good reason. Rivers provide easy access to water supply to meet urban water demands and a convenient transport route for trade and commerce. In modern times, urban riverfronts can be a great asset, potentially providing a pleasant contrast to the built environment of the urban landscape and a place for relaxation and reflection. Particularly in developing countries, however, poor river health and neglect of riverfronts have compromised these advantages, where degraded water quality makes river water unsuitable for human use and riparian corridors are unattractive and abandoned. Conditions are aggravated by inadequate sewerage systems and wastewater treatment, poor solid waste management (SWM), reduced river flows because of upstream water use and land management, and many other issues. Far from being an asset to be embraced by the city, the city turns its back on the river.

2. Of the top 10 most polluted large rivers in the world, 9 are in the Asia and Pacific region, of which 6 are in the People's Republic of China (PRC).[1] The important factors in the assessment of river health are (i) water quality; (ii) water quantity, with regard to availability of flow for environmental requirements and seasonal flow distribution; and (iii) environmental condition of the river corridor, including aquatic and riparian ecosystems. Without adequate water of sufficient quality to sustain biota critical to the ecological food web, the ecosystems could collapse, resulting in rivers that become ugly polluted drains virtually devoid of life; there are numerous examples of such rivers throughout Asia.[2] River rehabilitation aims to restore river health and create river corridors that provide attractive open space for community use and recreation, which are an asset for urban living.

3. Along river valleys, rivers connect urban and rural districts. Groundwater bodies also connect urban and rural districts, and they are connected to the river and its tributaries. Point sources of water pollution occur in both urban and rural areas; but, in quantitative terms, the worst point sources of pollution are typically urban or industrial in origin. Nonpoint sources (NPSs) of pollution, including agricultural runoff loaded with fertilizers and pesticides, and discharges from intensive animal husbandry, generally have more impact on water quality in rural areas. Reduced river flows are brought about by climate change, changing land use, and increasing demand for water, which aggravates the deterioration of water quality (less water but more pollutants), especially during dry spells. Hence, integrated urban–rural management of water resources is essential for the improvement of river health, i.e., management strategies that tackle sources of pollution in both urban and rural areas, and aim to maintain adequate river flows and seasonal distribution of river flows for environmental purposes.

4. Urbanization increases impermeable surfaces and reduces stormwater infiltration. Urban stormwater drainage systems cause local inundation whenever drain capacities are exceeded. By delivering more runoff more rapidly to receiving streams, stormwater drains aggravate downstream flood hazards, including in rural

---

1 The following are the world's 10 most polluted large rivers: Citarum (Indonesia); Yangtze (PRC); Indus (Pakistan); Yellow (PRC); Hai (PRC); Ganges (India); Pearl (PRC); Amur (PRC–Russian Federation border); Niger (Benin, Guinea, Mali, Niger, and Nigeria); and Mekong (Cambodia, Lao People's Democratic Republic, Myanmar, PRC, Thailand, and Viet Nam).

2 United Nations Environment Programme. 2016. *A Snapshot of the World's Water Quality: Towards a Global Assessment*. Nairobi.

areas. There is also accelerated instability and erosion of riverbanks, and urban encroachment in riparian zones with disposal of construction waste and riverbed aggradation. Urban stormwater drainage also transports pollutants to streams from sources like untreated urban waste and wash-off of residues and sediments.

5.   There is a vital interdependency between urban and rural areas. With respect to water resources, integrated management is required for the joint interests of urban and rural dwellers; the most appropriate scale for integrating water resources management is the river basin. This report derives from the Henan Dengzhou Integrated River Restoration and Ecological Protection Project, an Asian Development Bank (ADB) loan project in Henan Province in the PRC.[3] It is intended to rehabilitate and restore the badly degraded Tuan River in Dengzhou City. The strategy that guided project formulation attempts to address the important linkages between the management of water resources in the urban and rural districts of Dengzhou City, embracing interventions in multiple sectors of river basin water resources management and related environmental protection. This is referred to as the urban–rural water linkages (URWLs) approach. Acknowledging the urban–rural linkages, project activities include rural water supply, urban and rural wastewater treatment, NPS pollution control, flood risk management, urban drainage and stormwater management, wetland construction, sewer pipeline construction, SWM, urban park development, amenity space greening, heritage protection, works to rehabilitate river reaches that link urban and rural zones, and institutional capacity development. The project is useful as a model to demonstrate how these linkages can be addressed in other parts of the PRC or elsewhere to restore river health and prevent river assets from becoming eyesores.

B.   Global and Regional Water Challenges

6.   Many countries face similar issues. The world is becoming increasingly urbanized. An estimated 55% of the world's population resides in urban areas and this percentage is projected to reach 68% by 2050.[4] In the PRC, the urbanization rate is expected to reach 70% by 2035.[5] This urbanization poses major challenges for water managers. People need to be provided with adequate water and sanitation services, and be protected against water-related hazards (e.g., floods, degraded water quality). Major investments are needed to achieve acceptable levels of water security.[6] Food security is also a major concern, and agriculture is the greatest consumer of water resources in Asia (Box 1). The effects of climate change make these resource problems even more challenging.

7.   In 2015, the United Nations adopted the 2030 Agenda for Sustainable Development and its associated Sustainable Development Goals (SDGs).[7] The SDGs provide a universal and integrated guide for countries to eradicate poverty and achieve sustainable development globally by 2030. Its core content encompasses 17 goals and 169 specific targets covering the 3 aspects of sustainable development—economy, social affairs, and environment. SDG 6 (ensure availability and sustainable management of water and sanitation for all) is specifically oriented to water.

8.   Urban–rural linkages in water resources management are relevant to several SDGs and many associated targets. The main water-related goal (SDG 6) has specific targets for achieving safe and affordable drinking water supply (6.1), achieving adequate and safe sanitation and hygiene (6.2), reducing pollution (6.3), implementing integrated water resources management (IWRM) (6.5), and protecting and restoring water-related ecosystems (6.6). Other SDGs and targets address water-related objectives, with the primary

---

3   ADB. PRC: Henan Dengzhou Integrated River Restoration and Ecological Protection Project.
4   United Nations, Department of Economic and Social Affairs, Population Division. 2019. *World Urbanization Prospects 2018: Highlights*. New York.
5   Data were collected from the National Academy of Economic Strategy under the Chinese Academy of Social Sciences.
6   ADB. 2016. *Asian Water Development Outlook 2016: Strengthening Water Security in Asia and the Pacific*. Manila.
7   United Nations. 2015. *Transforming Our World: The 2030 Agenda for Sustainable Development*. New York.

Box 1: Water Security in the Asian Region

Asia is home to half of the world's poorest people. Its population is expected to reach 5.2 billion by 2050 and its megacities to increase to 22 by 2030. Given such growth, the region's finite water resources will be put under immense pressure. Recent estimates suggest that, by 2050, approximately 3.4 billion people could be residing in Asia's water-stressed areas. In addition, a staggering 1.7 billion people lack access to basic sanitation.

As a result of escalating demand from the industry and domestic sectors, the region's water demand is predicted to grow by 55% by 2050. The agriculture sector will, therefore, need to produce more food (60% more worldwide and 100% more in developing countries), utilizing the dwindling water resources. Water for agriculture continues to encompass 80% of the region's water resources consumption. These challenges are exacerbated by increasing water-related disasters and climate variability. From 1995 to 2015, a total of 2,495 water-related disasters hit the Asian region, resulting in 332,000 deaths and affecting 3.7 billion more individuals.

In general, the region reveals a widening gap in household water security—both between rich and poor, and between urban and rural areas. Poor governance and weak institutional capacity remain as major obstacles to enhancing water security.

Source: Asian Development Bank. 2016. *Asian Water Development Outlook 2016: Strengthening Water Security in Asia and the Pacific*. Manila.

focus on improving people's living conditions. Examples are targets for combating waterborne diseases (3.3); reducing economic losses because of natural disasters, with a focus on protecting the poor in urban areas (11.5); achieving the sustainable management and efficient use of natural resources (12.2); strengthening resilience and adaptive capacity to climate-related hazards (13.1); and protecting, restoring, and promoting the sustainable use of terrestrial ecosystems and their services (15.1).

9.   Looking at prevailing issues and challenges in the Asia and Pacific region, ADB has formulated its Strategy 2030, which promotes integrated solutions that move away from a silo approach in supporting sector development to a more holistic cross-sector approach.[8] Strategy 2030's operational plans call for upscaling best practices and adopting emerging innovative solutions to meet the region's water challenges. Out of Strategy 2030's seven operational priorities, operational priority 3 includes (i) enhancing efforts to address water quality, water productivity, and IWRM; (ii) strengthening the linkages between water, food, and energy; and (iii) enabling integrated solutions.[9] For operational priority 4, holistic solutions are key, including smart water management, a sponge city approach for urban flood management,[10] nature-based solutions, and inclusive sanitation to cover both networked and non-networked wastewater management systems.[11] Under operational priority 5, ensuring food security will require improved productivity through modernization of irrigation systems, including automated control systems, water reuse, micro-irrigation, and pressurized-pipe irrigation. Remote sensing, use of drones, and smartphone applications need upscaling to improve irrigation efficiency.[12] In fact, water is a cross-sector issue in all operational priorities in Strategy 2030.

---

8   ADB. 2018. *Strategy 2030: Achieving a Prosperous, Inclusive, Resilient, and Sustainable Asia and the Pacific*. Manila.
9   ADB. 2019. *Strategy 2030 Operational Plan for Priority 3: Tackling Climate Change, Building Climate and Disaster Resilience, and Enhancing Environmental Sustainability, 2019–2024*. Manila.
10 The sponge city approach is "a sustainable urban planning and design approach aimed at reducing stormwater (urban flood) and polluted runoff, and at reusing water in ecologically friendly ways. It builds a water system infrastructure that acts like a sponge to absorb excess rainfall and surface runoff, store and purify rainwater, and release it for reuse. Design techniques include permeable surfaces, gardens, rainwater harvesting, green spaces, and lakes." R. Osti. 2018. Integrating Flood and Environmental Risk Management: Principles and Practices. *ADB East Asia Working Paper Series*. No. 15. Manila: ADB. p. 15.
11 ADB. 2019. *Strategy 2030 Operational Plan for Priority 4: Making Cities More Livable, 2019–2024*. Manila.
12 ADB. 2019. *Strategy 2030 Operational Plan for Priority 5: Promoting Rural Development and Food Security, 2019–2024*. Manila.

10.   With the outbreak of the coronavirus disease (COVID-19), the world faces an unprecedented risk that is disrupting the lives, health, and well-being of multitudes. The COVID-19 global pandemic has brought about devastating economic and social consequences, as well as the loss of many lives worldwide. It has also exposed the fragile interdependency and balance between human society and the natural environment, which have been habitually taken for granted. Uncertainty looms as to how the post-pandemic world will emerge from this crisis; what is certain is that life as it was will never be the same. An important lesson to be learned from COVID-19 is that Mother Nature should be respected. Post-pandemic development efforts, therefore, need to incorporate sufficient harmony between humans and nature, with a focus on sustainability. Sound management of natural resources, including the water sector, should be promoted. Some of the possible scenarios in the water sector include the following: (i) water services (urban and rural) may suffer the most because of fiscal budget deficiencies, as government efforts will focus more on economic recovery; therefore, asset management will be a challenge; (ii) more financing for waste management in cities at the county level is needed to address rapid increases in industrial pollution, agricultural intensification, and climate change adaptation; (iii) the capacity of infrastructure may not be sufficient as more waste production or more water extraction for food production leads to more pollution of already strained resources; therefore, natural resources conservation, most importantly water and river health, will be at risk; (iv) business continuity management, mainly the continuity of water operations, is a major challenge during the COVID-19 pandemic for water utilities, including impacts on field operations and interruptions of waste treatment chemical supply chains; therefore, public utilities need to fundamentally change their operations to meet increased demand during the crisis; (v) integrated disaster risk reduction has become a prominent issue, and investments need to look at disaster risk management capacity more broadly, leading to the establishment of risk-financing mechanisms and platforms for integrated disaster risk reduction investments; and (vi) the institutional and governance dimensions of the water sector may need to be reviewed and enhanced by learning from the ongoing crisis.

C.   Water Resources Issues in the People's Republic of China

11.   Despite huge investment and significant policy initiatives since early 2000, major issues remain in the PRC with respect to the quantity and quality of surface water and groundwater. These water resources management challenges must continue to be addressed if the country is to achieve its socioeconomic and environmental development objectives. Increased water demands, in combination with low water use efficiency, have resulted in water shortages in many regions. Development levels and utilization of water resources are uneven across the country. Water shortage is more prevalent in the north and northwest, and less so in the south. Only 18% of water resources available nationally are in the northern half of the country (i.e., north of the Yangtze River Basin). Rural areas account for more than 60% of total water use, mainly for agriculture. Service delivery for domestic and industrial water use in developed urban regions is much higher than in the relatively backward rural regions.[13]

12.   In the PRC, the total annual fresh water available is about 2,800 million cubic meters (mcm), on average, of which about 822 mcm is from groundwater. Average annual per capita water availability is about 2,000 cubic meters (m^3), just one-third of the global average. Not all available water can be captured for use, and some of the water is so polluted it is unsuitable for use. Available water is also declining because of climatic factors related to global warming. Accompanying rapid economic growth since the PRC began to open up and reform its economy in 1978, agricultural production had to increase dramatically, and use of chemical

---

13   ADB. 2018. *Managing Water Resources for Sustainable Socioeconomic Development: A Country Water Assessment for the People's Republic of China.* Manila.

fertilizers and pesticides increased to almost double the recommended limits by international standards, e.g., chemical fertilizer applied at 492.6 kilograms per hectare. Because of urbanization, encroachment into more than 2,000 square kilometers (km^2) of ecological zones (river corridors, lake shores) has occurred since 2010. Soil erosion is critical in more than 2.95 million km^2 of land. Almost 80% of rural waste is directly discharged into water bodies because of a lack of wastewater treatment facilities (footnote 13). Although more than 83,000 urban wastewater treatment plants (WWTPs) have been established in the PRC, with a combined capacity exceeding 250 mcm per day, their utilization has not yet reached 50% because of various factors, including incomplete collection networks, improper locations, and continuing rapid expansion of urban areas (footnote 13). Therefore, a large portion of urban and industrial wastes is still being discharged into rivers untreated. As a result, 30% of main river reaches in the PRC have water quality worse than class IV,[14] more than 60% of groundwater is highly polluted, and over 77% of rivers and lakes experience eutrophication.[15] The situation is illustrated in Figure 1.

13.   Because of untreated pollutant discharges, water quality is a major challenge for the country. Dedicated pursuit of economic development to improve living standards has produced unmatched rates of socioeconomic progress, but too often at the expense of the environment and water-dependent ecosystems. This follows similar patterns in developed countries during earlier periods of rapid expansion and industrialization; however, in the PRC, it has been more dramatic precisely because the rate of development has been more rapid. Attention is now being turned to remedying those adverse environmental effects, including the deterioration of the conditions of the nation's lakes and rivers (Box 2). In the larger rivers of the PRC, close to 30% of river reaches have water quality worse than class IV. The situation in small and medium-sized rivers is generally worse, with flows during drier parts of the year greatly diminished. Groundwater pollution continues to be a prominent issue in many regions. It has spread from urban areas to peri-urban areas, where groundwater is often used as a source of drinking water.

14.   Water scarcity, which is most severe in the northern part of the country, is a major challenge for the improvement of river health and the water quality of its surface water and groundwater resources. Nationally, average annual water diversion and use in 2014 was about 600 billion cubic meters (bcm), which was short of the estimated water demand by about 17 bcm. According to the PRC's 5-year economic and social development plans, and allowing for improvements in water use efficiency and demand management initiatives, demand will increase to about 700 bcm by 2030 (footnote 13). Clearly, greater exploitation of the national water resources will be essential in the next decade, and it follows logically that less water will be available for environmental requirements and the restoration of river health. Figure 2 shows the short-term projections of water supply and demand. Note that use of national statistics ignores regional differences; hence, for some regions, the prognosis is even worse. Flows in some of the main rivers in the PRC, such as the Yellow River, already show a declining trend, and future effects of climate change will only make it more difficult to modify that trend (Figure 3).

15.   The Government of the PRC embarked on a series of reforms to address technical and institutional water-related challenges. The government gives priority to sustainable development and has initiated implementation of the 2030 Agenda for Sustainable Development. In September 2016, the government

---

14   According to the PRC's Environmental Quality Standards for Surface Water (GB3838-2002) prepared by the Ministry of Environmental Protection (now the Ministry of Ecology and Environment), water bodies (such as rivers, lakes, channels, and reservoirs) are categorized into five classes based on environmental functions and protection objectives: (i) class I primarily applies to water from the source (i.e., headwaters) and the national nature reserves; (ii) class II applies to the first-class protected areas for centralized water sources for living and drinking needs, the protected areas for rare aquatic organisms, and the feeding and spawning areas of fish and shrimps; (iii) class III is for the second-class protected areas for centralized water sources for living and drinking needs, protected areas for aquaculture, and swimming areas; (iv) class IV refers to water areas suitable for industrial (e.g., factory) use and noncontact recreation; and (v) class V is suitable only for agriculture use (irrigation) and landscaping. Water quality deteriorates from class I to class V, with water quality under class IV or class V designated as unsuitable for human water supply use.
15   Eutrophication is the process by which excessive nutrients accumulate in a water body, causing dense plant growth and death of aquatic life from lack of dissolved oxygen.

Figure 1: Some Key Water Resources Issues in the People's Republic of China

Fertilizers/area
492.6 kg/ha[a]

83,227 WWTPs
250 mcm/day
(<50% used)

Total Diversion or Use
609.5 bcm
(81% from surface water)

Annual Freshwater Resources
2,800 bcm

Surface Water
2,700 bcm

Groundwater
821.8 bcm

North
2,300 bcm

South
500 bcm

30% of river reaches class IV or worse[b]

37% of groundwater class I–III[b] (i.e., 63% highly polluted)

Polluted rivers and 77% of 121 large lakes eutrophic

Built-up area
50,000 km² (+64% in 10 years)

Reduced flow
From climatic factors and/or water withdrawal

Active soil erosion
2.95 million km²

bcm = billion cubic meters, ha = hectare, kg = kilogram, km² = square kilometer, mcm = million cubic meters, WWTP = wastewater treatment plant.

a   This amount is double the international standard.
b   Under the national Environmental Quality Standards for Surface Water (GB3838-2002), water quality deteriorates from class I to class V. Water quality of class IV or class V is unsuitable for human water supply use.

Source: Asian Development Bank.

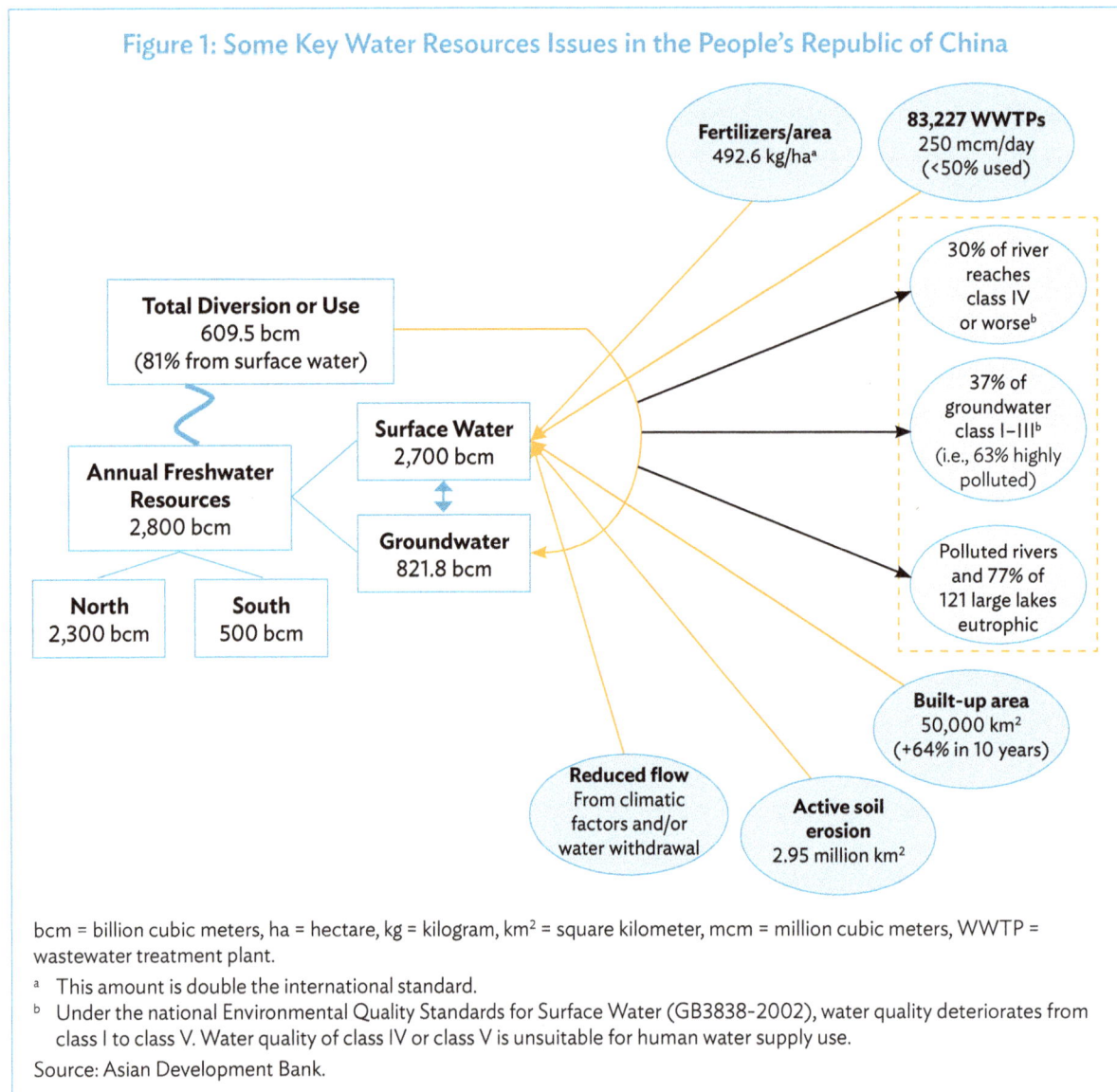

adopted the National Plan on Implementation of the 2030 Agenda for Sustainable Development, which includes guidance and general principles for the implementation of the 17 SDGs.[16]

16.   Many highly advanced as well as conventional water technologies have been used to address river water pollution considering both point sources and NPSs of pollution in the PRC. The PRC is perhaps the most proactive developing country in terms of formulating national policies for environmental protection (Chapter III). The government has made great efforts to improve the water quality in its rivers. For example, the 12th and 13th five-year plans covering 2011–2020 placed even stronger focus (in comparison to the previous plans) on green growth and river pollution reduction, with significant parts of national budgets expended on environmental improvement. There was also significant progress made during the 10th and 11th five-year plans covering 2001–2010. However, the Government of the PRC has taken numerous initiatives

---

16   Government of the PRC, Ministry of Foreign Affairs. 2016. *China's National Plan on Implementation of the 2030 Agenda for Sustainable Development*. Beijing.

Box 2: Water Security in the Yellow River Basin

Overextraction and other climatic and human factors led to the drying of large sections of the lower Yellow River for over 200 successive days at some points in the past. Therefore, the State Council of the People's Republic of China formulated the Yellow River Water Allocation Scheme in 1987, with clear water entitlements for 11 provinces in the river basin. However, this scheme encountered implementation and monitoring difficulties, resulting in little progress in addressing the water crisis in the Yellow River Basin over the next 15 years. In 2002, the transboundary flow requirement and real-time monitoring provisions were endorsed by the State Council, which contributed to some improvements in the implementation of the scheme. However, many other issues emerged, such as river sedimentation and pollution, making water security in the basin more complex and challenging.

Sources: G. Pegram et al. 2013. *River Basin Planning: Principles, Procedures and Approaches for Strategic Basin Planning.* Paris: United Nations Educational, Scientific and Cultural Organization; and Government of the People's Republic of China, Ministry of Water Resources, General Institute of Water Resources and Hydropower Planning and Design. 1987. *Yellow River Water Allocation Scheme.* Beijing.

Figure 2: Projections of National Water Supply and Demand

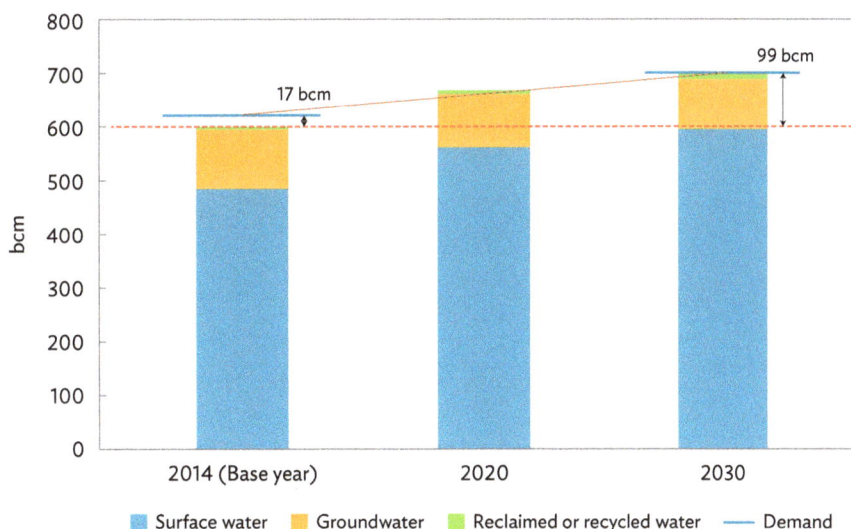

bcm = billion cubic meters.
Source: Asian Development Bank. 2018. *Managing Water Resources for Sustainable Socioeconomic Development: A Country Water Assessment for the People's Republic of China.* Manila.

since early 2000 that sometimes conflict with one another, resulting in duplicated or confused objectives among local governments. Too many policies and plans can obscure key priorities; thus, without firm guidance and direction, local governments may be unsure how to properly implement programs that are necessary. There was improvement in water quality earlier this century, particularly during the period of the 10th and 11th five-year plans (2001–2010) (Figure 4). But, despite several recent reforms, improvement has stalled since 2012. River health remains a major problem and future progress is uncertain mainly because of an increasing water demand and supply gap, climatic variability, and unprecedented urbanization (paras. 11–13). The government should consider trying an alternative approach to address water resources problems.

Figure 3: River Basin Flow Changes, 1980–2014

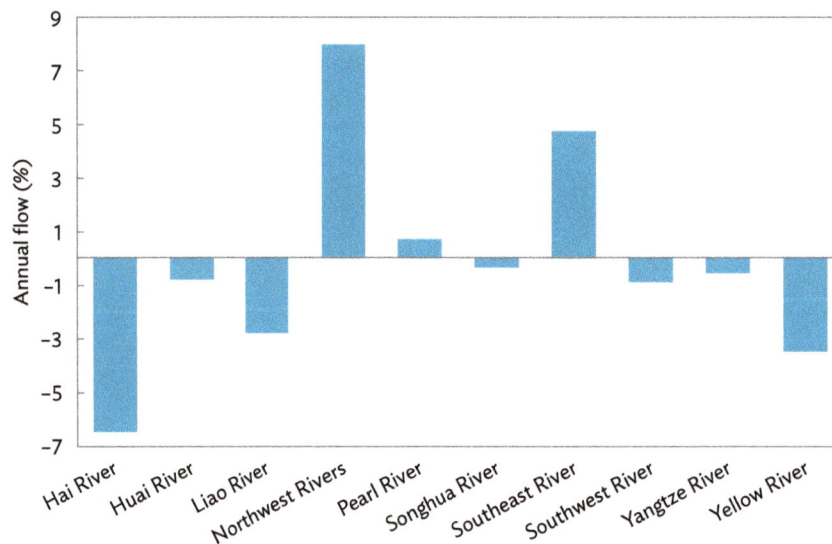

Source: Asian Development Bank. 2018. *Managing Water Resources for Sustainable Socioeconomic Development: A Country Water Assessment for the People's Republic of China*. Manila.

Figure 4: Water Quality Trend in the Main Rivers of the People's Republic of China, 2003–2014

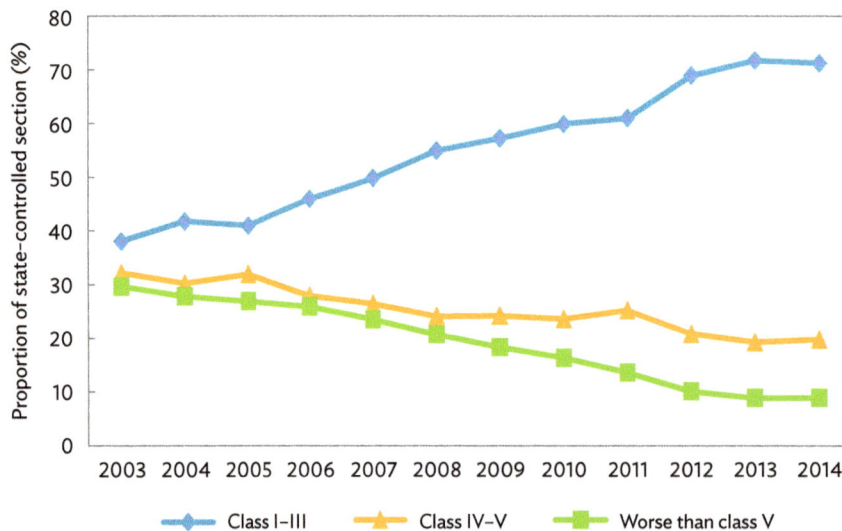

Note: According to the Environmental Quality Standards for Surface Water formulated by the Ministry of Environmental Protection (now the Ministry of Ecology and Environment) of the People's Republic of China, water bodies (e.g., rivers, lakes, and reservoirs) are divided into five classes according to utilization purposes and protection objectives: (i) class I is mainly applicable to water from the source (headwaters) and the national nature reserves; (ii) class II applies to the first-class protected areas for centralized sources of drinking water, the protected areas for rare fish, and the spawning fields of fish and shrimps; (iii) class III is for the second-class protected areas for centralized sources of drinking water, protected areas for common fish, and swimming areas; (iv) class IV refers to water areas suitable for industrial use and entertainment not directly touched by human bodies; and (v) class V is suitable only for irrigation and landscaping.

Source: Asian Development Bank. 2018. *Managing Water Resources for Sustainable Socioeconomic Development: A Country Water Assessment for the People's Republic of China*. Manila.

17.   The PRC also suffers from frequent flood and drought disasters each year, which constrain socioeconomic development in both urban and rural areas. In the past, river and flood management in the PRC traditionally applied hard engineering measures to try to control nature. Toward the end of the 20th century, the Government of the PRC began shifting its attention to greater use of nonstructural measures and nature-based approaches (green management measures) in conjunction with hard (gray) engineering measures. This mode of more integrated management can build resilience and provide greater cost-effectiveness in managing flood and environmental risk. Nonstructural and green management measures require lower installation and maintenance costs. Green management measures confer flexibility that allows ecosystems to adjust autonomously to a range of climatic conditions. While respecting the natural dynamics and flow of fresh water, sediments and nutrients, and people's dependencies on these at river basin scale, the green management approach aims to facilitate social and economic development in a more sustainable manner. It can guide optimal management and investment strategies for long-term sustainable river management and associated management objectives.

18.   River rehabilitation is not only about restoring the ecological functioning of the river system. It provides opportunities for social benefits (improved public health and increased social well-being and recreational opportunities in crowded cities) and economic benefits (reduced health-care costs, lower implementation costs, and other monetary and nonmarket benefits). These benefits accrue to both urban and rural areas within a river basin. However, past river rehabilitation practice is no longer effective because it failed to recognize the importance of URWLs, and this is perhaps the missing element in current PRC development.

19.   The Government of the PRC recognizes the need for an integrated approach to address these water management issues, but still seeks effective ways of achieving this in both urban–rural and regional developments. In the past, there was a focus on urban areas as they were central in the drive for rapid economic development and, therefore, received more political attention than rural areas. The development of urban water supply systems, which often rely on sources from the rural areas, often had adverse impacts on rural water supply. However, in the national interest, water is also vital for sustaining agricultural development, food security, and livelihoods in rural communities. More support from the government is needed now to promote water management and socioeconomic development in the rural areas, and not just for reasons of equity. An integrated approach of urban and rural development will benefit both urban and rural areas because the two are inextricably linked and their welfare is interdependent. The PRC has embarked on a series of reforms to address the technical and institutional water-related challenges (Chapter III).

D.   Urban–Rural Linkages and the Need for Integration

20.   There are no clear boundaries between urban and rural areas—the urban core seamlessly transitions to peri-urban and then rural areas, and there are physical connections linking the areas to one another, including water resources systems. The groundwater system underlies both urban and rural areas, while rivers, originating from rural areas, pass through or beside cities and continue to flow on to other urban and rural areas.

21.   The binding connection between urban and rural areas means that developments taking place in an urban area will often have impacts on rural areas. For example, urban pollution degrades water quality for rural areas. Conversely, land use and water use in rural areas impact cities. For example, soil erosion increases sediment transport in rivers causing sediment deposition downstream, and NPSs of fertilizers and pesticides from agricultural land use pollute water sources upon which urban water supply may rely. When river flows are reduced, their capacity to dilute and assimilate contaminants declines, and river health and environment deteriorate in both urban and rural reaches. Likewise, groundwater exploitation for irrigation can impact on dry weather river flow as there will be no reversal recharge of river from the groundwater, and the same can cause land subsidence in both urban and rural areas. Therefore, in river management and rehabilitation, important linkages exist between interventions in urban and rural areas. Proper planning and design of

water services in urban and rural areas will serve to better protect and conserve the river environment. Rehabilitating rivers in urban areas will benefit downstream rural areas. Integrated planning of interventions will improve outcomes for urban and rural areas, so an integrated urban–rural approach is needed to address these water linkages.

22.   The urban–rural linkages related to water are illustrated in Figure 5, using the situation of Dengzhou in the Tuan River Basin as an example. In the figure, the urban (upstream) area is on the right and the rural (downstream) area is on the left. Local governments often use rivers as a reference for subnational administrative boundaries, including between urban and rural areas. In doing so, different local administrative units often ignore URWLs, and there is a lack of coordination and spatial planning that results in fragmented approaches and plans. However, since URWLs often span administrative boundaries, urban and rural areas (perhaps unwittingly) influence each other in terms of water quantity (withdrawal and availability) and water quality (pollution) of both surface water and groundwater.

23.   The issues for water resources management (surface water and groundwater) are often regarded as problems of too much, too little, or too dirty. The causes of these problems are generally found in urban and rural areas, and good water resources management requires integrated interventions in both. Clearly, water quality in a major river can only be improved if the pollution from both urban and rural areas is reduced.

24.   In the broader social and environmental context, neglect of URWLs has direct implications for urban and rural livelihoods, and for the physical environment that sustains urban and rural communities. For example, untreated sewage and solid wastes from the urban area (on the right of the river in Figure 5) is discharged directly into the river, which recharges groundwater that is the only source of drinking water in the rural area (on the left of the river in Figure 5). Similarly, poor sanitation and SWM in the rural villages, untreated discharges from animal husbandry such as piggeries, and the use of chemical fertilizer and pesticides on the farmlands contaminate the groundwater and the river water that meet both urban and rural demand. Polluted river water and soil erosion from inappropriate rural land use degrade the river environment and the riverfront corridor in the urban area. Urban encroachment and disposal of construction wastes into the river contaminate water in the river and reduce its conveyance capacity, leading to more frequent flooding both locally and downstream.

25.   Three categories of physical URWLs that drive these water management issues in a river basin may be distinguished: (i) the use of land in the basin (leading to soil erosion and diffuse pollution), (ii) the use and management of the water resources (affecting availability of surface water and groundwater), and (iii) water pollution (degrading quality of surface water and groundwater). These drivers link the urban and rural areas through the river and groundwater systems (Figure 6).

26.   The use of land influences water-related processes in the watershed. Changes in land use (e.g., urbanization, deforestation) modify the water balance, seasonal characteristics of runoff, the sediment balance (by erosion, sediment transport, and deposition), water quality (by point sources and NPSs of pollution), groundwater quantity and quality, and flooding (urban and rural). Water resources and water availability are also modified by hydraulic structures (e.g., dams, weirs, intake structures). The most important urban–rural linkages regarding land use are as follows:

   (i)   **From rural to urban.** Changes in land use (e.g., deforestation, agricultural practices) will modify the hydrological regime (low and high flows) and soil erosion (leading to higher sediment load and higher nutrient levels in rivers and streams). While higher flows can cause more flooding, lower flows will impact the availability of water for consumers and the ecosystems (environmental water requirements).

   (ii)   **From urban to rural.** Urbanization leads to changes in runoff characteristics. Urban developments generally increase runoff, and stormwater drainage leads to more rapid runoff. However, there are

Figure 5: Schematic of Identified Urban–Rural Water Linkages, Tuan River Basin

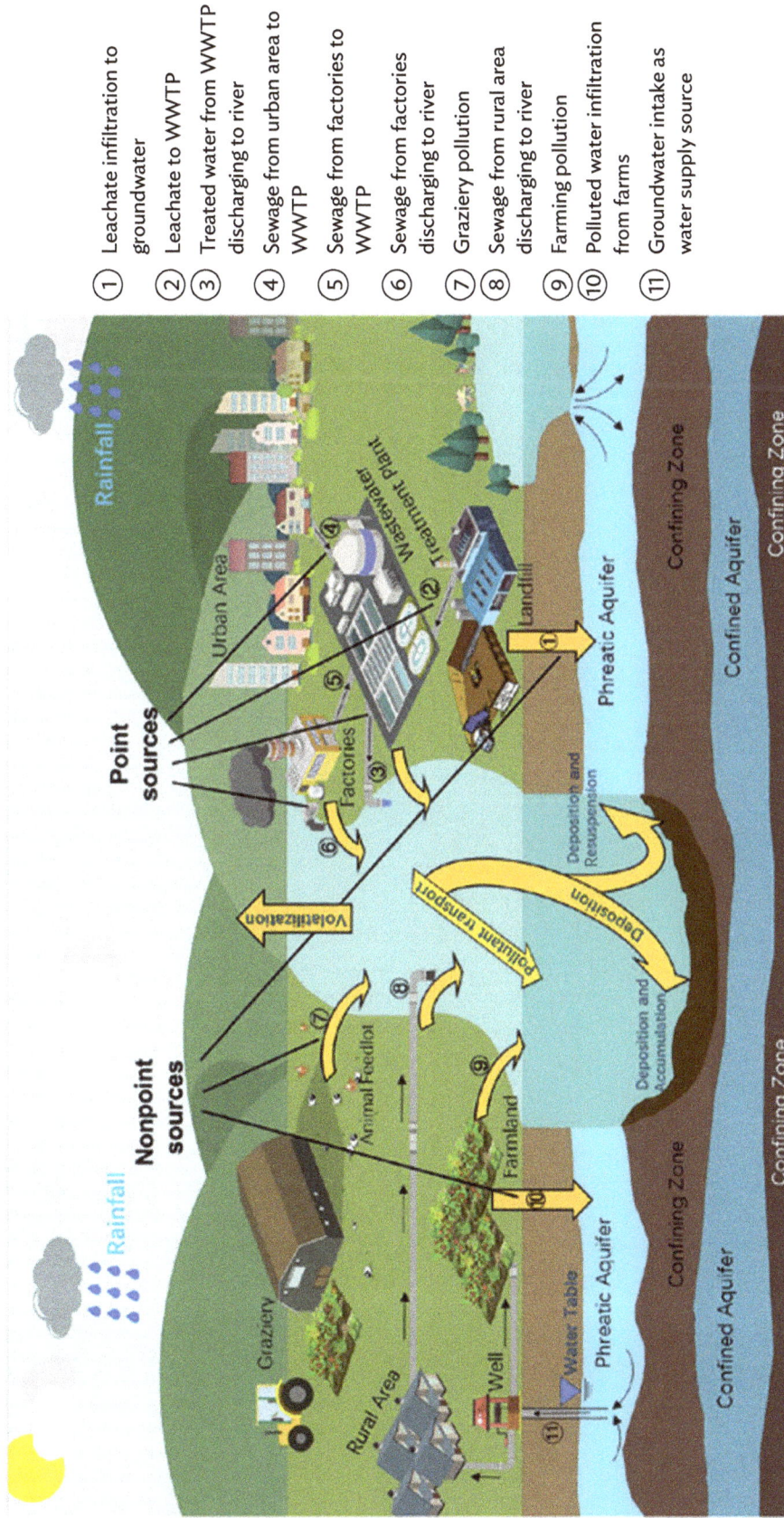

1. Leachate infiltration to groundwater
2. Leachate to WWTP
3. Treated water from WWTP discharging to river
4. Sewage from urban area to WWTP
5. Sewage from factories to WWTP
6. Sewage from factories discharging to river
7. Graziery pollution
8. Sewage from rural area discharging to river
9. Farming pollution
10. Polluted water infiltration from farms
11. Groundwater intake as water supply source

WWTP = wastewater treatment plant.
Source: Asian Development Bank.

Figure 6: Categories of Influences of the Urban–Rural Water Linkages

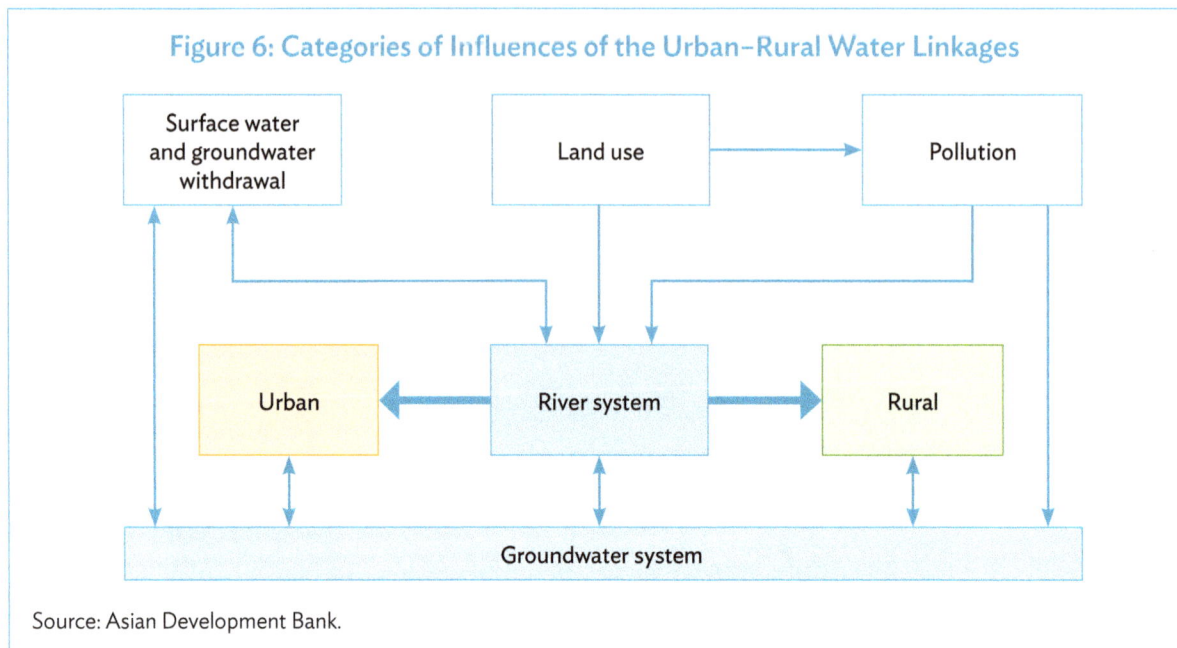

Source: Asian Development Bank.

methods for slowing down runoff (temporary storage) or maintaining infiltration to groundwater (e.g., the sponge city approach in the PRC, known by other names in different parts of the world).

27.   The withdrawal of water by water users makes less water available for other users; however, part of the water diverted for use will become available again when it is returned as drainage water (polluted in urban areas, surplus to demand in irrigation areas). This applies to both surface water and groundwater. Urban–rural linkages regarding water withdrawals include the following:

(i)   **From rural to urban.** Withdrawal of water upstream (e.g., for irrigated agriculture) can constrain the water available for domestic and industrial water supply in urban areas. Groundwater withdrawal in a large volume for irrigation may be the key reason for land subsidence in urban areas. In coastal or estuarine zones, it may even lead to salinity intrusion, endangering water intakes of downstream users.

(ii)   **From urban to rural.** Withdrawals for urban use reduce availability of water for downstream rural users. These include withdrawals for interbasin water transfer.

28.   Water use and land use generally lead to pollution of surface water or groundwater systems—pollution that constrains the use of water by others and may have economic and public health impacts. SWM is also important in combating water pollution. Even in communities with organized solid waste collection and disposal, toxic leachates from waste disposal facilities (e.g., landfills, incineration plants, composting sites) may contaminate surface water or groundwater. The strategic location of landfill sites as well as WWTPs—therefore, spatial planning—has impacted water quality in the river basin. The main urban–rural linkages on pollution are as follows:

(i)   **From rural to urban.** Poor water supply and sanitation practices in rural areas degrade the water quality of surface water and groundwater on which others rely. Pollutants from agricultural activities (fertilizers, pesticides, animal manure if untreated) can lead to serious deterioration of surface water and groundwater quality. High levels of nutrients in surface water can lead to eutrophication and algal blooms.

(ii)   **From urban to rural.** Urban domestic and industrial wastewater degrades the water quality of surface water and groundwater. Some pollutants (such as nutrients, organics, and heavy metals) accumulate in the bed sediments of rivers. Both direct and indirect users of the river water will be at risk. Pollution of groundwater by urban sources may threaten the health of rural communities.

29.   Despite these important urban–rural linkages, in many developing countries, the provision of water supply and sanitation services for urban and rural residents is quite separate. In the PRC, for example, service providers for urban water supply charge for service, typically using tiered tariff schemes based on water used. Although full cost recovery remains an aspiration, at least the charges for urban water are usually adequate to cover operational costs.[17] In the rural areas of the PRC, including small rural communities, no fees or minimal fees are charged for water supply as of yet.[18] As a consequence, service delivery in rural areas is inferior, often unreliable, and can be a risk to public health. Water-related ailments are more prevalent in rural areas. Government policy initiatives for integrated urban–rural development and for rural vitalization will focus attention on improving socioeconomic conditions in rural areas, and are intended to narrow the gap between services provided for urban and rural residents.

30.   There is very strong social and economic interdependence between urban and rural areas. Short- or long-term migration from rural areas to urban areas provides labor for the urban workforce and economy, and alleviates rural poverty. Urban centers depend on the food (e.g., fresh vegetables), fiber, and other resources from nearby rural areas. Water and the water-related urban–rural linkages play an important role in the rural production of this food and fiber. Water is an essential input to production, but the pollution of water resources, which occurs in both urban and rural areas, is a major public health concern throughout river basins—in particular, the adverse impact of waterborne diseases. The social and economic impacts of poor waterfront design in the urban areas because of poor water quality and a degraded river environment are huge and long lasting.

---

17   D. Tan. 2014. Pricing Water. *China Water Risk.* 13 January.
18   R. Rutkowski. 2014. The Economics of H2O: Water Price Reforms in China. *China Economic Watch.* Peterson Institute for International Economics. 22 July.

II

Urban–Rural Water Linkages in Tuan River, Dengzhou City

A. Dengzhou City

31.   Dengzhou City is in southwestern Henan Province. Its area of jurisdiction is 2,369 square kilometers (km²), with a population in 2018 of 1.78 million, of which 60% are rural residents. With respect to water resources management, Dengzhou City is of strategic significance as it is in the zone of the South-to-North Water Diversion Project (SNWDP), a major national water management project in the People's Republic of China (PRC). The SNWDP diverts water from Danjiangkou Reservoir on the Han River via a 1,400-kilometer (km) canal (basin transfer) to more than 30 cities in the water-scarce and drought-prone northern region of the PRC, including Beijing, Tianjin, and other cities in the provinces of Hebei and Henan (Map 1).

32.   In Dengzhou City, the urban core seamlessly transitions to peri-urban and then rural areas (Map 2). The urban core continues to expand, with new town centers emerging around the periphery every few years. As the city expands, the peri-urban areas evolve, and rural areas retreat. From 2011 to 2018, the urban area within Dengzhou City expanded from about 20 km² to approximately 32 km².[19] According to the city master plan, urban areas will expand to approximately 90 km² by 2030.

33.   Almost all rural households and 30% of urban households in Dengzhou City do not have proper access to water supply, wastewater treatment, or solid waste management (SWM) services. Most still rely on groundwater as their source of water supply, posing a significant public health risk, as the groundwater has been badly contaminated and cases of waterborne diseases have been elevated since 2015.[20]

B. Tuan River Basin

34.   Dengzhou City is located within the Tuan River Basin and within Henan Province, PRC (Map 1). Jituan station is the most downstream hydrological station, near where the Tuan River joins the Zhao River. The total watershed area at Jituan station is about 5,000 km². Further downstream, the Zhao River joins the Han River, a primary tributary of the Yangtze River.

C. Land Use Impacts

1. The Physical Processes

35.   **Hydrology.** Average annual precipitation in Dengzhou is about 800 millimeters, of which 66% occurs in the wet season from June to September. The annual average river discharge at Jituan station is about 76 million cubic meters (mcm), with about 80% of runoff occurring during wet season months. The river is very badly

---

19   Footnote 3, estimate by project consultants based on Google map images.
20   According to the Dengzhou rural water supply survey and the feasibility study report, 480 of the 588 villages in the rural areas have excessive fluorine and 232 have excessive nitrite. Recorded cases of waterborne diseases in these areas have reached 292,000 since 2015.

Map 1: Project Location—Dengzhou City in Nanyang City of Henan Province

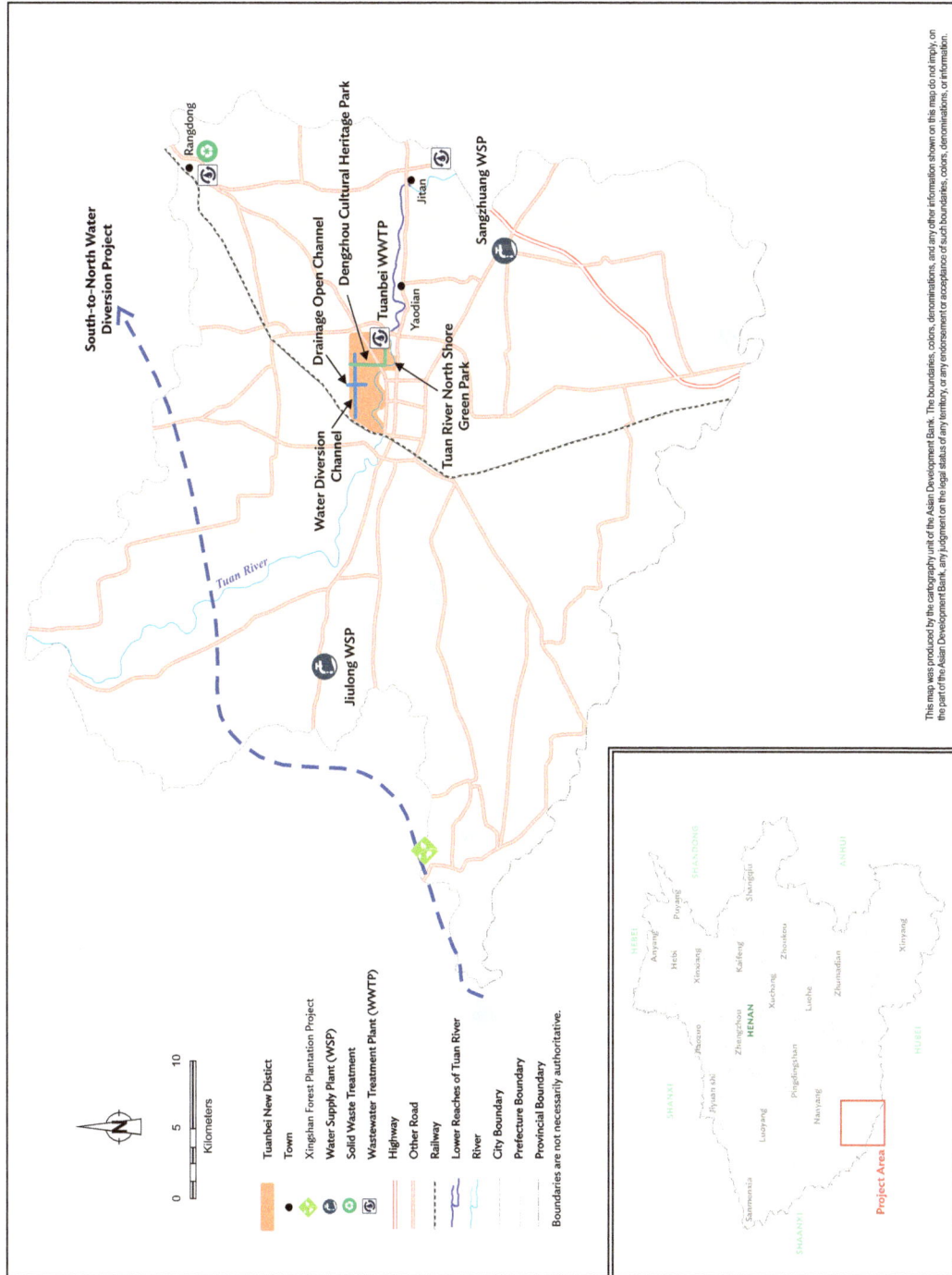

South-to-North Water Diversion Project

Rangdong

Dengzhou Cultural Heritage Park

Drainage Open Channel

Tuanbei WWTP

Jitan

Sangzhuang WSP

Yaodian

Water Diversion Channel

Tuan River North Shore Green Park

Tuan River

Jiulong WSP

Legend

- Tuanbei New Distict
- Town
- Xinghan Forest Plantation Project
- Water Supply Plant (WSP)
- Solid Waste Treatment
- Wastewater Treatment Plant (WWTP)
- Highway
- Other Road
- Railway
- Lower Reaches of Tuan River
- River
- City Boundary
- Prefecture Boundary
- Provincial Boundary
- Boundaries are not necessarily authoritative.

Kilometers
0   5   10

N

Inset map

HEBEI

Anyang

Puyang

SHANXI

Hebi

Xinxiang

SHANDONG

Jiaozuo

Zhengzhou

Kaifeng

Shangqiu

HENAN

Sanmenxia

Luoyang

Xuchang

Zhoukou

ANHUI

Jiyuan shi

Pingdingshan

Luohe

Zhumadian

Nanyang

Xinyang

SHAANXI

HUBEI

Project Area

This map was produced by the cartography unit of the Asian Development Bank. The boundaries, colors, denominations, and any other information shown on this map do not imply, on the part of the Asian Development Bank, any judgment on the legal status of any territory, or any endorsement or acceptance of such boundaries, colors, denominations, or information.

20-0679a AV

Note: The prefecture-level city of Nanyang in Henan Province, People's Republic of China administers the county-level city of Dengzhou.

Source: Asian Development Bank.

Map 2: Aerial View of Dengzhou City

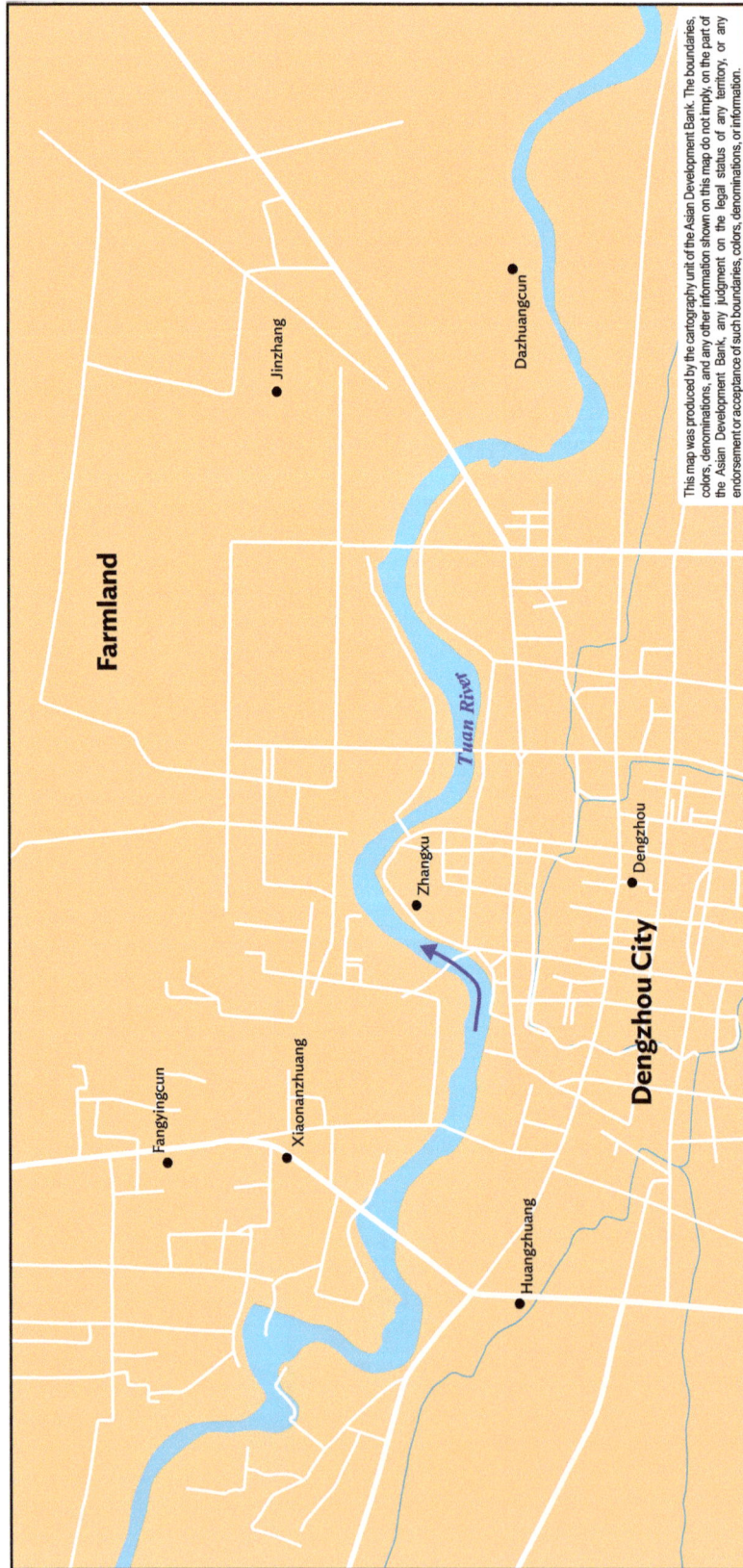

Farmland

Tuan River

Jinzhang

Dazhuangcun

Fangyingcun

Xiaonanzhuang

Zhangxu

Huangzhuang

Dengzhou

Dengzhou City

This map was produced by the cartography unit of the Asian Development Bank. The boundaries, colors, denominations, and any other information shown on this map do not imply, on the part of the Asian Development Bank, any judgment on the legal status of any territory, or any endorsement or acceptance of such boundaries, colors, denominations, or information.

20-0679b AV

Note: Farmlands on the north bank and urban area on the south bank of the Tuan River.

Source: Asian Development Bank.

polluted, worse than the class V standard along some of its length (footnote 14). Through the urban reach of the river and downstream, the river water is eutrophic and has a persistent foul odor.

36.   **Surface–groundwater interconnection.** Groundwater recharge occurs by infiltration and percolation from the topsoil surface as well as from the Tuan riverbed. Contamination of groundwater and river systems takes place from (i) landfills, (ii) septic systems, (iii) leaky underground storage tanks, (iv) fertilizers and pesticides applied to agricultural land, (v) industrial facilities, and (vi) road salts and other chemical wash-off. Pollutants are also conveyed to rivers and streams by drains and ditches, and direct sewage discharge from unsewered areas contributes to river pollution problems. Groundwater moves through aquifers to discharge areas, particularly the Tuan River during dry periods, or is extracted for domestic and other use throughout the year. Overexploitation of water resources in the river basin and climate change have reduced river flows significantly, especially in the dry season; therefore, the capacity of rivers to dilute and assimilate contaminants is reduced, increasing the severity of water pollution.

37.   **Flooding.** Flooding is a natural phenomenon, but is aggravated by changed land use in urban and rural areas. Typical examples that increase volumes of runoff and lead to accelerated runoff are deforestation in rural areas and reduced infiltration capacity and stormwater drainage in urban areas. Substantial construction wastes from Dengzhou's urban area have been dumped in the river, resulting in reduced conveyance capacity of the river channel in the city and significant riverbed aggradation in downstream areas. Results of river hydraulic modeling indicate that sediment deposition in the Tuan River reduces the flood conveyance capacity by approximately 5%–10%. As a result, the frequency of flooding in urban and rural areas of Dengzhou City is increasing, particularly in the rural villages about 30 km downstream of the city. Urban area expansion in flood-prone areas has not only reduced the flood retention capacity but has also changed the flood footprints in the basin.[21]

Flooding in Dengzhou City. During a storm event in May 2018, floods inundated both rural (left) and urban (right) areas of Dengzhou City (photos from the Dengzhou City Government).

21   R. Osti. 2019. *Institutional and Governance Dimensions of Flood Risk Management: A Flood Footprint and Accountability Mechanism.* *ADB East Asia Working Paper Series.* No. 24. Manila: ADB.

38.   **Erosion and sedimentation.** The soil erosion rate in some rural areas in Dengzhou City alone is more than 100 tons per hectare (ha) per year, which is beyond the threshold of 50 tons per ha per year for high erosion as defined by the Food and Agriculture Organization of the United Nations.[22] Throughout the year, runoff transports suspended solids and nonpoint source (NPS) pollutants—e.g., total nitrogen, total phosphorus, and organic pollutants—into the rivers and streams of the Tuan River Basin. Analysis indicates that the majority of suspended solids and phosphorus loads occur during the wet season. Suspended solids flux exceeds 4,600 tons per day during the wet season but is only 24 tons per day during the dry season. Most of the Tuan River Basin consists of farmland and undeveloped land, with a small portion of low-density residential areas and villages. The landscape conditions are susceptible to soil erosion. Nutrients wash off the land from NPSs of pollutants (mainly agricultural land) and are transported with suspended solids into receiving waters.

39.   Together with the adverse effects of water pollution, sediments deposited in rivers degrade aquatic habitats and food webs. They also have caused major reductions in fish numbers and serious disruptions of freshwater ecosystems. Nutrients can activate cyanobacteria to release toxins from bed deposits hazardous to human health.

40.   Located at the head of the SNWDP (and along its middle route), the Dengzhou City Government (DCG) faces development restrictions imposed mainly because water quality and quantity in the Danjiangkou Reservoir is being degraded, in part by runoff from within Dengzhou City. To ensure sound environmental management and protection of the reservoir, the DCG receives eco-compensation annually from the Beijing Municipal Government as an incentive; however, this has not been used well.[23]

2.   Spatial Planning

41.   Spatial planning plays a critical role in water pollution. For example, the location of the waste treatment facility should be consistent with the overall land use plan of the city. There are two wastewater treatment plants (WWTPs) in the Dengzhou urban area, with total treatment capacity of 60,000 cubic meters (m³) per day, which is far more than the required capacity; wastewater production in Dengzhou City is roughly 44,000 m³ per day, but total treatment ratio is only about 50% (para. 51). The Sanda WWTP in the southeastern part of the city has 127 km of sewer network, but that has not covered the major part of the city, where urbanization is taking place. The topography and various social factors restrict the coverage of WWTPs. The landfill sites and WWTPs are not strategically located, not just because of their coverage, but also because their proximity to water bodies allows leachate water to directly contaminate the groundwater and river water in the upstream of the urban area.

3.   Urban–Rural Linkages in Land Use

42.   The discussion in paras. 36–41 demonstrates the importance of interconnections between land use in urban and rural areas and the linkages that occur through the water systems. Developments in land use in rural areas, such as deforestation and unsustainable agricultural practices, lead to more variability in water availability (causing more frequent floods and droughts, and low dry season flows), deterioration in water quality, and increased sediment loads. On the other hand, untreated wastewater from urban areas discharged to rivers and

---

22   Food and Agriculture Organization of the United Nations and Intergovernmental Technical Panel on Soils. 2015. Chapter 6: Global Soil Status, Processes, and Trends. *Status of the World's Soil Resources—Main Report*. Rome.

23   The eco-compensation received by the DCG from Beijing includes CNY278 million in direct cash payments and CNY340 million for capacity building in environmental protection, including medical coverage and education for affected populations. However, the amounts are part of the DCG's annual budget, and there is no proper accounting provision for the compensation fund by source and area. The eco-compensation package also includes free use of 692 mcm of water from the SNWDP for various uses—only about 15% of this has been used by the DCG because of lack of proper planning and limited water resources management capacity.

infiltrating groundwater systems impacts the rural areas in terms of their ability to use that water for drinking and other domestic purposes.

43.   Since 2000, the natural drainage to the Tuan River Basin has changed, with the disappearance of many small creeks and ditches in the landscape because of land use change and encroachment arising from agricultural development and urbanization. This alteration to the landscape has transformed the drainage and discharge of runoff to the Tuan River, exacerbating local flooding and adversely affecting groundwater by causing increased infiltration and/or percolation and the transport of pollutants to shallow groundwater aquifers.

D.   Water Withdrawal and Water Use

1.   Drinking Water Supply

44.   With the improvement of living standards, demand for water has been increasing in both urban and rural areas of Dengzhou City. The urban water supply is serviced by water treatment plants. A centralized groundwater drinking well, with a capacity of 30,000 m³ per day, is the main source of urban water supply, providing up to 54% of drinking water used by urban residents. The remainder is supplied from shallow groundwater wells. Rural residents do not have access to reticulated water supply systems or clean surface water, leading to use of shallow groundwater as their main source of drinking water. In general, the shallow groundwater underlying both urban and rural areas is

Water supply from groundwater. This is a typical water well used in the rural areas of Dengzhou City (photo from the Dengzhou City Government).

contaminated by recharge from the highly polluted Tuan River, by fertilizers and pesticides, and because of poor sanitation and SWM. Since 2008, almost 300,000 cases of waterborne diseases have been reported. Most occurred in rural areas where there was a higher incidence of morbidity among women and children. Overexploitation of the groundwater is a related issue, and is unsustainable in the long term. The problems arising from the disturbance to the water cycle are complex. During the dry season, lowered water tables because of overpumping constrain water supply and increase costs of extraction in urban and rural areas, and also induce recharge from the badly polluted rivers to which they are connected. In the wet season, however, higher groundwater tables can occur because of poor land drainage, which can increase the rate of release of polluted groundwater to the Tuan River and its tributaries; river or groundwater recharge varies spatially and temporally.

2.   Irrigation

45.   Henan Province has a large portion of farmland, where 52% of irrigated lands are served by shallow tube wells mostly constructed by individual farmers. Groundwater overextraction results in water table decline in many areas, including in Dengzhou City. The overextraction of groundwater, increases in groundwater pumping costs, and the degradation of surface water and groundwater quality from agricultural runoff and infiltration are the key issues.

3.   Instream Flow Needs (Environmental Water Requirements)

46.   Healthy rivers are a vital part of the natural environment. In addition to providing habitat for fish and wildlife, rivers contribute to scenic and aesthetic qualities of natural settings, offer opportunities for recreation, and provide water for livestock and other uses. Minimum flows are required instream to sustain a healthy environment all along the river channel, in both urban and rural reaches. Meeting these requirements should be a key constraint on upstream water use and groundwater withdrawals, but past developments in Dengzhou and the Tuan River Basin have largely ignored these requirements to the detriment of river health. In response to major reduction in dry season river flows, the DCG constructed small dams in the river channel to pool the water in an attempt to improve the urban aesthetics. It also protected the riverbank by constructing a concrete wall. These reactive measures, however, have instead induced deterioration in the river condition (e.g., putrid water in the pooled urban reach emits a foul odor), contributed to sediment deposition, and aggravated flooding. Basin-scale water resources management plans, along with effective programs to implement them, are urgently needed to adequately maintain low flow conditions in the Tuan River during dry seasons.

Tuan River during the dry season. The low flow conditions in the Tuan River during dry seasons should be maintained through effective water resources management planning (photos from the Dengzhou City Government).

E.   Pollution

47.   Serious pollution of the Tuan River is considered a priority water management issue for Dengzhou City. The river water quality is classified class IV or class V at different locations, indicating water quality so poor that it is not suitable for any urban or rural use (footnote 14). The health of the Tuan River is directly linked to the conditions and use of the land that drains into it. Among the major sources of pollutants (Figure 7 and Figure 8) that impair the Tuan River water quality, NPS pollution is the major culprit. Locations of pollution sources of major rivers and tributaries in the Tuan River Basin are presented in Map 3.

1.   Nonpoint Source Pollution

48.   Unlike pollution from industrial and sewage treatment plants, NPS pollution derives from many diffuse sources. Most NPS pollution derives from agricultural land use and is transported via surface runoff or land drainage from farmland. According to the 2015 land use analysis within the Tuan River Basin, farmland and pasture together account for 71.2% of land use, while urban areas occupy only 1.0% (Table 1).[24] Even within the limits of Dengzhou City, over 60% of land use is farmland.

---

24   DCG. 2019. *Initial Environmental Examination (Draft): Henan Dengzhou Integrated River Restoration and Ecological Protection Project in the PRC* (prepared for ADB).

Figure 7: Pollutant Analysis of the Tuan River

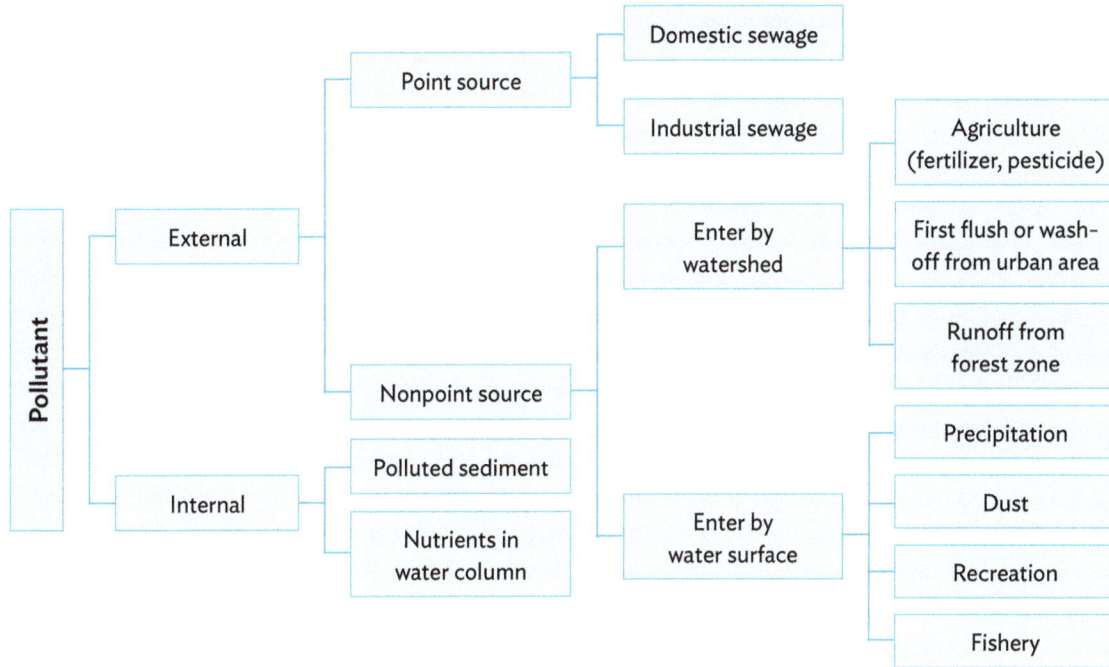

Source: Asian Development Bank.

Figure 8: Illustration of Point and Nonpoint Sources of Pollution

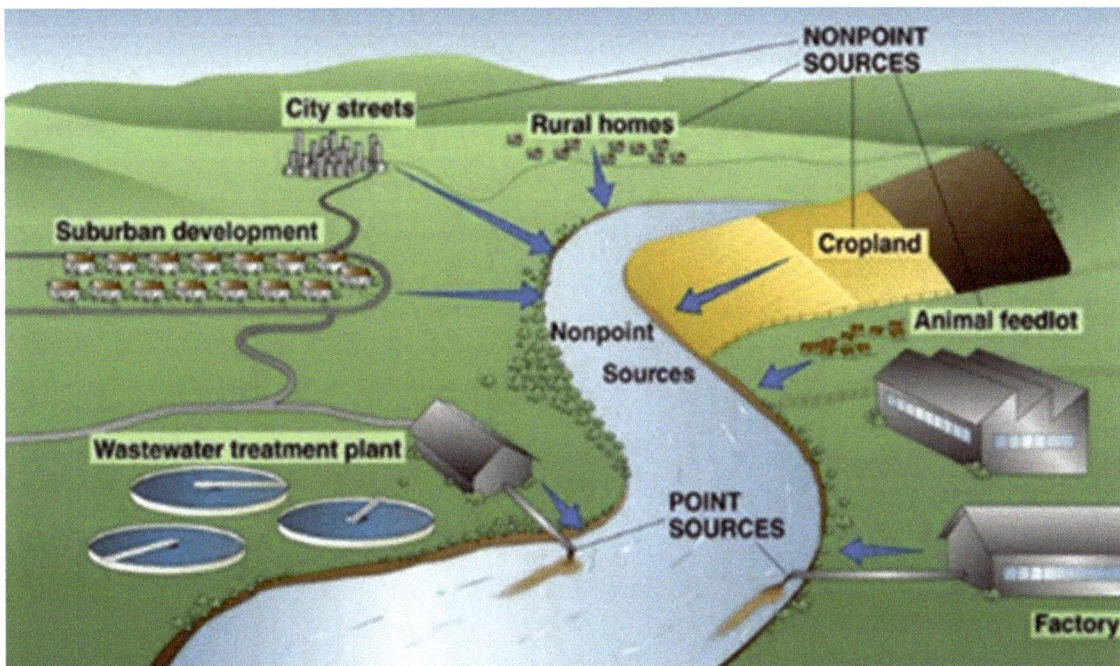

Source: Dengzhou City Government. 2019. *Initial Environmental Examination (Draft): Henan Dengzhou Integrated River Restoration and Ecological Protection Project in the People's Republic of China* (prepared for the Asian Development Bank). Reprint of Figure V-30, p. 134.

Map 3: Locations of Key Point and Nonpoint Sources of Pollution in the Tuan River Basin

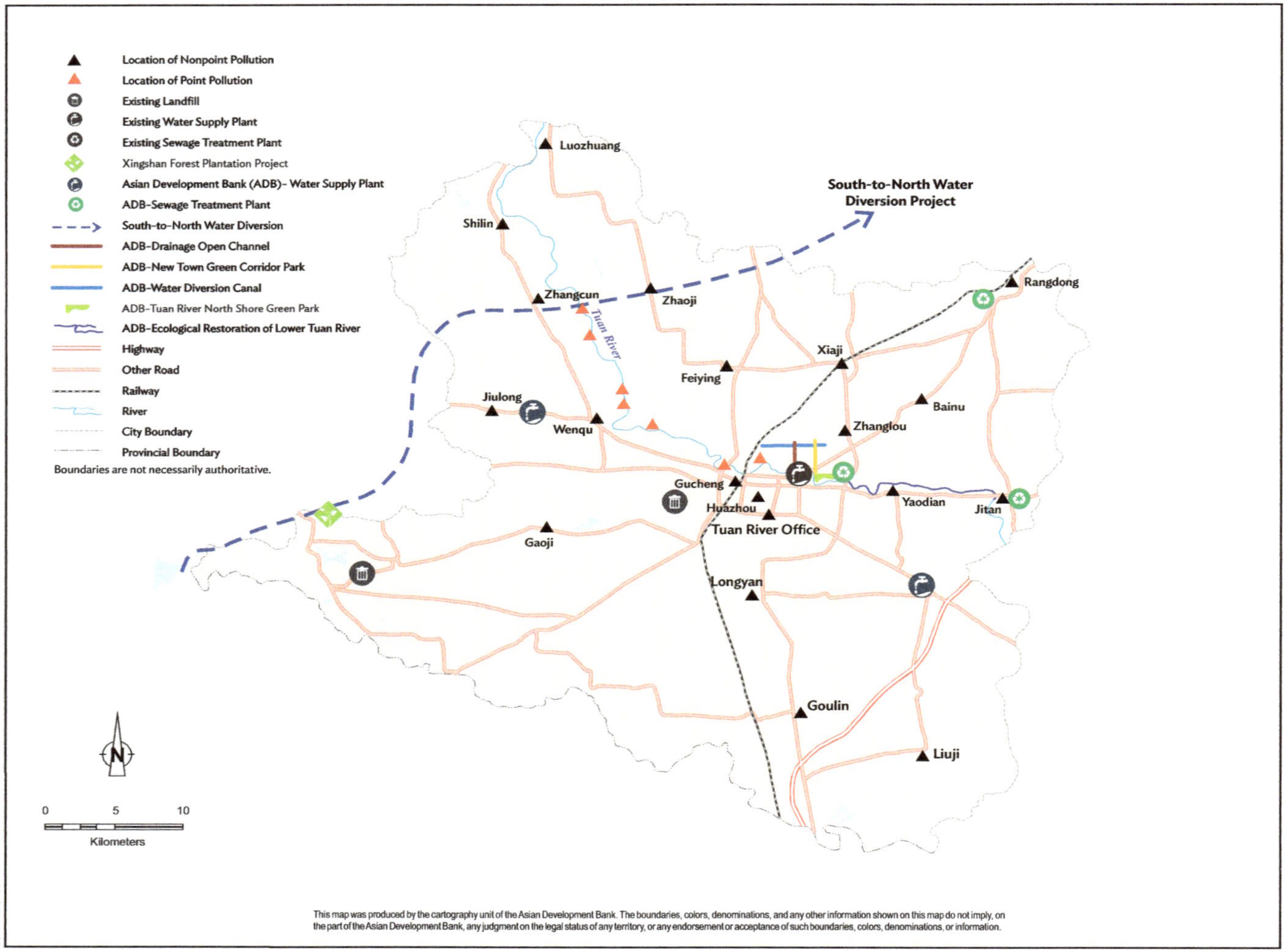

Legend

- Location of Nonpoint Pollution
- Location of Point Pollution
- Existing Landfill
- Existing Water Supply Plant
- Existing Sewage Treatment Plant
- Xinghan Forest Plantation Project
- Asian Development Bank (ADB)–Water Supply Plant
- ADB–Sewage Treatment Plant
- South-to-North Water Diversion
- ADB–Drainage Open Channel
- ADB–New Town Green Corridor Park
- ADB–Water Diversion Canal
- ADB–Tuan River North Shore Green Park
- ADB–Ecological Restoration of Lower Tuan River
- Highway
- Other Road
- Railway
- River
- City Boundary
- Provincial Boundary

Boundaries are not necessarily authoritative.

South-to-North Water Diversion Project

Rangdong
Jitan
Bainu
Zhanglou
Yaodian
Liuji
Xiaji
Goulin
Longyan
Tuan River Office
Feiying
Huazhou
Gucheng
Zhaoji
Luozhuang
Tuan River
Zhangcun
Wenqu
Gaoji
Shilin
Jiulong

0   5   10
Kilometers

This map was produced by the cartography unit of the Asian Development Bank. The boundaries, colors, denominations, and any other information shown on this map do not imply, on the part of the Asian Development Bank, any judgment on the legal status of any territory, or any endorsement or acceptance of such boundaries, colors, denominations, or information.

20-0679c AV

Source: Asian Development Bank.

Table 1: Land Use in the Tuan River Basin, 2015

Land Cover	Area (km²)	Area (%)
Agriculture	3,046	60.9
Forestry	1,235	24.7
Pasture	515	10.3
Range brush	131	2.6
Water	26	0.5
Urban	49	1.0

km² = square kilometer.

Source: Dengzhou City Government. 2019. *Initial Environmental Examination (Draft): Henan Dengzhou Integrated River Restoration and Ecological Protection Project in the People's Republic of China* (prepared for the Asian Development Bank).

Wastes from animal husbandry. Animal wastes are becoming major sources of water pollution (photos from the Dengzhou City Government).

49.   The average amount of fertilizer use in the city area is 615 kilograms (kg) per ha, much higher than the recommended safe dosage of 225 kg per ha.[25] Pesticide use is approximately 190 tons per year. The estimated fertilizer loss into the Tuan River is 18,000 to 21,100 tons per year.[26] It is estimated that NPS pollution contributes approximately 75% of the total phosphorus load in the river.

50.   Animal manure is a major source of nitrogen and phosphorus in surface water and groundwater. The rapid increase in the number of livestock has resulted in more grazing in cropland and pastures, and more concentrated animal feeding operations.

---

25 Xichuan County Government. 2018. *Project Proposal Document*. Xichuan.
26 Footnote 3, project report by the transactional technical assistance consultants.

Wastewater discharges to the Tuan River. Wastewater is directly discharged into the Tuan River system through various outfalls (photos from the Dengzhou City Government).

2.   Point Source Pollution

51.   **Discharge of untreated wastewater.** The amount of wastewater generated from Dengzhou City is 44,000 m³ per day, of which 39,000 m³ is from domestic sewage and the other 5,000 m³ from industrial wastewater discharge. There are two municipal WWTPs in Dengzhou City, with total combined capacity of 60,000 m³ per day. However, the city has utilized only about 50% of the available WWTP capacity because of several factors (para. 12). The WWTPs discharge treated effluent to the Tuan River or local tributaries. Other smaller, privately owned WWTPs are operated by large industrial water users, yet none of their effluents meet the quality standards.

52.   There are still urban areas unconnected to sewerage networks, and sewage from unserviced areas is discharged directly into rivers without treatment, some illegally. According to a survey by the Dengzhou Environmental Protection Bureau, approximately 3.3 mcm of untreated sewage is directly discharged into the Tuan River system each year.

53.   The city now also provides sewerage service to peri-urban areas on the north bank of the Tuan River—i.e., the Tuanbei New District, which the DCG intends to develop into a new urban area with quality infrastructure and services (paras. 102–108). The city has extended sewerage service to these areas by collecting and pumping sewage across the Tuan River to existing WWTPs, but this is unsustainable because of the high cost and the potential threat of pollution to the river. Therefore, a new WWTP is proposed in the Tuanbei New District.

54.   The DCG has given waste discharge permits to 24 industrial enterprises in the Tuan River Basin. The annual chemical oxygen demand load from these industrial discharges is estimated to be 935 tons, and the ammonia nitrogen (NH_3-N) load is 583 tons. In a local tributary of the Tuan River, the largest pollutant load is generated by a paper mill operated by Yixin Enterprise. Although the mill effluent meets the minimum national pollutant discharge standards, the large volume of its effluent discharge (1 mcm per day) has a substantial impact on the Tuan River water quality.[27] The DCG has required the paper mill to upgrade its effluent standard to improve water quality in the Tuan River.

---

27   DCG, Dengzhou Environmental Protection Bureau. 2017. *Implementation Plan of Dengzhou City's Atmospheric and Water Pollution Prevention and Control Campaign in 2017* (in Chinese). 19 April.

3.   Solid Waste Management

55.   Though not always dealt with in water management programs, insufficient SWM is a major source of water pollution. Over time, wastes dumped in landfills decompose and release concentrated toxic liquids. These leachates are highly hazardous. Companies who manage modern landfills have effective drainage systems to collect leachates and store them in sealed containers, which can be taken away for specialized safe disposal. Otherwise, even small amounts of leachate can badly contaminate surface water and aquifers, rendering them unsuitable for human use. In 2017, Dengzhou City produced almost 1,500 tons of garbage, kitchen wastes, and sewage sludge daily.[28]

56.   In rural and peri-urban areas of the Tuan River Basin, some villages have no garbage collection tanks, and few are equipped with collection bins. Garbage is not sorted for recycling. Community collection tanks are often not emptied on time, so overflow of rubbish is common. Consequently, garbage is discarded on roadsides; on open spaces around houses; on hillsides; and in depressions, drains, ponds, and riverbanks. This not only has an adverse visual impact on the living environment and countryside, but also presents a health risk for local residents.

57.   **Groundwater contamination from urban landfills.** There are two urban sanitary landfills with a combined design capacity of 360 tons per day; in 2017, they were accepting about 480 tons per day. These two landfill sites are reaching their capacity limit far ahead of their design period because of the sharp increase in volume of solid waste from the urban area. Clearly, Dengzhou City does not have sufficient landfill capacity to handle all the garbage it produces. The DCG has proposed a new SWM facility in the Vein Industrial Park Development Plan that will greatly improve future SWM.[29]

58.   **Groundwater contamination from rural landfills.** There are six temporary garbage storage sites in the rural parts of Dengzhou City. Because of the inadequate leachate collection and disposal measures, it is estimated that these rural sites produce nearly 100,000 tons of leachates annually. These untreated toxic leachates infiltrate into the groundwater of the Tuan River Basin, poisoning the surrounding environment. These rural temporary garbage storage sites are partly surrounded by wheat and croplands; therefore, there is potential to contaminate the crops, which is of serious concern as a public health risk to the food chain.

4.   Surface Water and Groundwater Quality in the Tuan River Basin

59.   Linkages between point sources and NPSs of pollution and the surface water and groundwater systems immediately downstream of the main urban center of Dengzhou are illustrated in Map 4. Waste discharges into the Tuan River from various sources—including livestock wastes, untreated sewage, fertilizers, pesticides, and solid waste leachates—degrade the quality of surface water and groundwater through runoff, drainage, infiltration, and recharge. Water quality in almost all sections of the lower Tuan River exceeds class V (the worst on the scale), indicating noncompliance with the class III national standard target.[30] Some of the tributaries of the Tuan River are also badly polluted.

---

28   ADB. 2019. *Report and Recommendation of the President to the Board of Directors: Proposed Loan to the PRC for the Henan Dengzhou Integrated River Restoration and Ecological Protection Project.* Sector Assessment (Summary): Agriculture, Natural Resources, and Rural Development (accessible from the list of linked documents in Appendix 2). Manila.

29   The provincial government of Henan has formulated a plan to construct the Vein Industrial Park by 2020, where incineration of approximately half of the province's garbage can be done. In addition to this, the DCG also plans to build a local treatment plant by 2025 to incinerate 95.0% of the city's garbage (footnote 28, para. 7).

30   ADB. 2019. *Report and Recommendation of the President to the Board of Directors: Proposed Loan to the PRC for the Henan Dengzhou Integrated River Restoration and Ecological Protection Project.* Manila. para. 6.

Map 4: Linkages between Waste Pollution and River Recharge in the Tuan River Downstream of Dengzhou

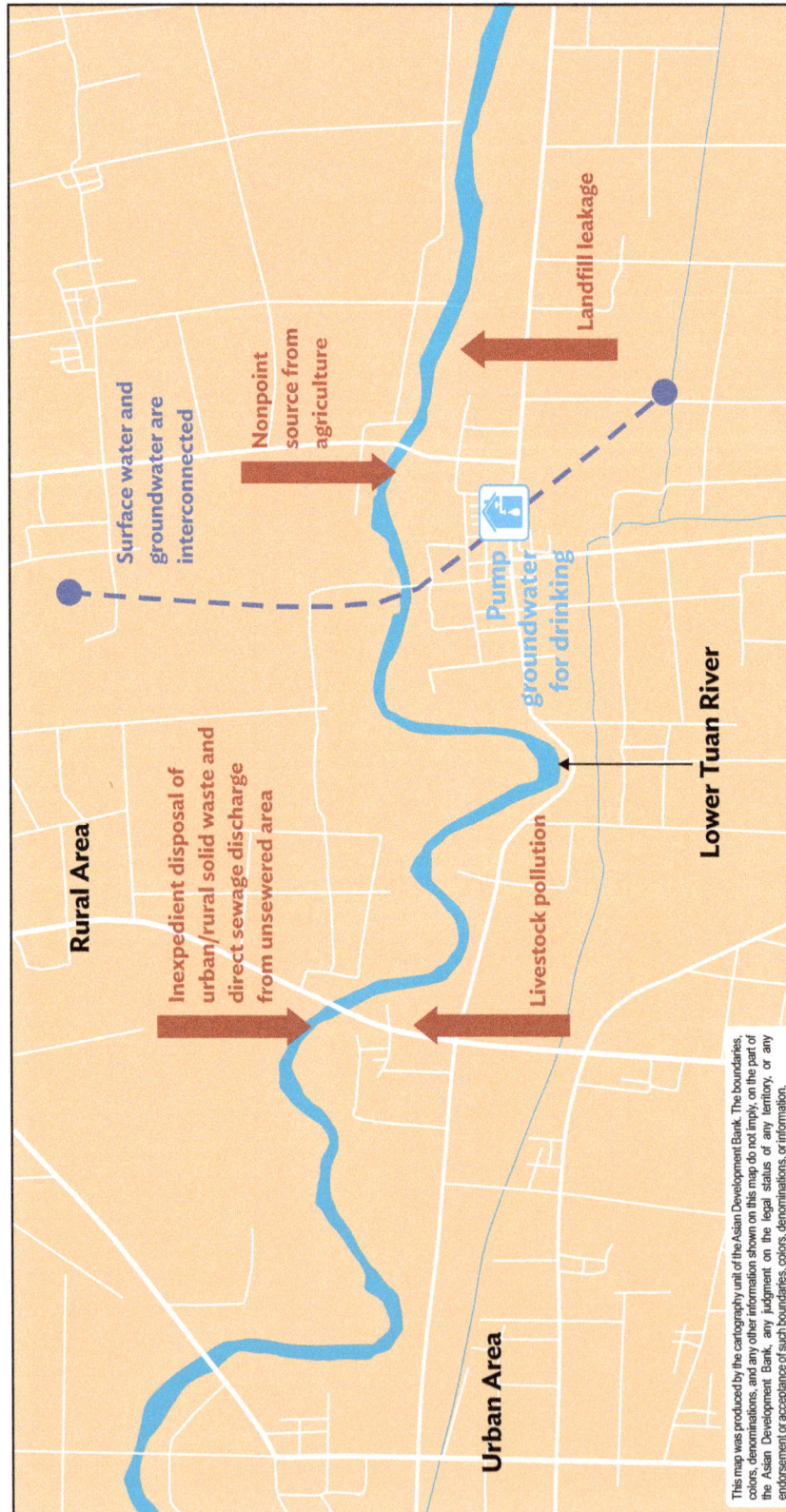

Rural Area

Urban Area

Lower Tuan River

Surface water and groundwater are interconnected

Nonpoint source from agriculture

Landfill leakage

Pump groundwater for drinking

Inexpedient disposal of urban/rural solid waste and direct sewage discharge from unsewered area

Livestock pollution

This map was produced by the cartography unit of the Asian Development Bank. The boundaries, colors, denominations, and any other information shown on this map do not imply, on the part of the Asian Development Bank, any judgment on the legal status of any territory, or any endorsement or acceptance of such boundaries, colors, denominations, or information.

20-0679d AV

Source: Asian Development Bank.

60.   Furthermore, substantial quantities of nutrients, toxic substances, and heavy metals are accumulating in the sediments of the Tuan River. Laboratory testing of bed material samples, conducted on 12 March 2019, revealed high concentrations of organic substances, total phosphorus, and total nitrogen in sediments of the lower Tuan River. During high flow events, pollutants in these contaminated sediments may be disturbed and transported through the water column because of scouring. Particulate organic matters deposited in the sediment bed may go through mineralization or decomposition processes, known as diagenesis, while dissolved inorganic nutrients in the sediment bed may be recycled back to the river as sediment flux. Nutrients released from the bed sediment can cause eutrophication (or algal growth issues).

61.   The water quality of groundwater in the Tuan River Basin is badly degraded, with concentrations of fluoride, total dissolved solids, and nitrite in rural villages far exceeding the national drinking water standards (GB5749-2006).[31] More than 430,000 rural residents in Dengzhou City rely on groundwater for water supply, including drinking water. Groundwater quality in urban areas also fails to meet the national drinking water standards. Since 2008, about 300,000 cases of waterborne diseases were reported, mostly in rural areas.

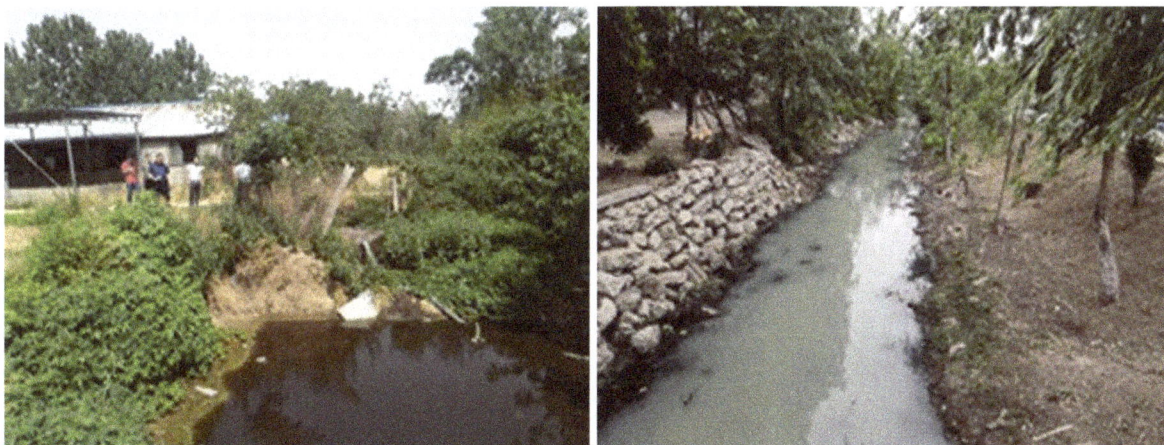

Rural wastewater discharges. Wastewater discharges in the rural areas of Dengzhou City come from various sources (photos from the Dengzhou City Government).

62.   In the rural parts of Dengzhou City, where the population has reached approximately 1.2 million, few towns and villages have wastewater treatment facilities. High loads of biochemical oxygen demand (BOD), total nitrogen, and total phosphorus are discharged into receiving waters.[32] Some of the wastewater is discharged directly into small creeks and rivers, some leaks from household septic tanks, and some is simply disposed of on the ground surface.

63.   Since groundwater is used as the main source of water supply in rural areas, waste discharges from urban areas also have great potential to affect the quality of drinking water for rural residents. As the groundwater and surface water systems are well interconnected, regardless of where the untreated sewage is discharged in the river basin, the pollutants will eventually enter the Tuan River and the groundwater. An integrated river restoration plan is needed for Dengzhou City, with due consideration of URWLs.

---

31   DCG. 2011. *Integrated Water Resources Planning of Dengzhou City (2011–2030)*. Dengzhou City.
32   The numbers were calculated with the assumptions of 25 grams of BOD per capita per day, 5 grams of total nitrogen per capita per day, and 0.7 grams of total phosphorus per capita per day—these are lower values, according to the Code for Design of Outdoor Wastewater Engineering, GB50014-2006 (2014).

5.   Food Supply Chain Impact and Public Health Risk

64.   The rural population in Dengzhou City relies strongly on water supply from groundwater for both irrigation and domestic use. However, their source of water has been contaminated because of recharge from polluted surface waters, untreated domestic wastes, and infiltration of runoff from farmlands where chemical fertilizers and pesticides are widely applied. There are about 2.6 million *mu* (173,333 ha) of cultivated land within the municipal boundary of Dengzhou City.[33] There is a risk that its irrigation with polluted surface water and groundwater could contaminate crops and vegetables for human consumption, i.e., contamination of the food supply chain.

65.   Figure 9 illustrates the potential impact of pollutant sources on the food supply chain, which could adversely affect the health of residents in urban and rural areas. Polluted domestic water sources also endanger the safety of food supply chains, as they may be used to clean fruit, vegetables, and other food produce. The polluted Tuan River has posed further public health risk because of people's direct contact with the polluted water through washing clothes, swimming, and bathing in the river.

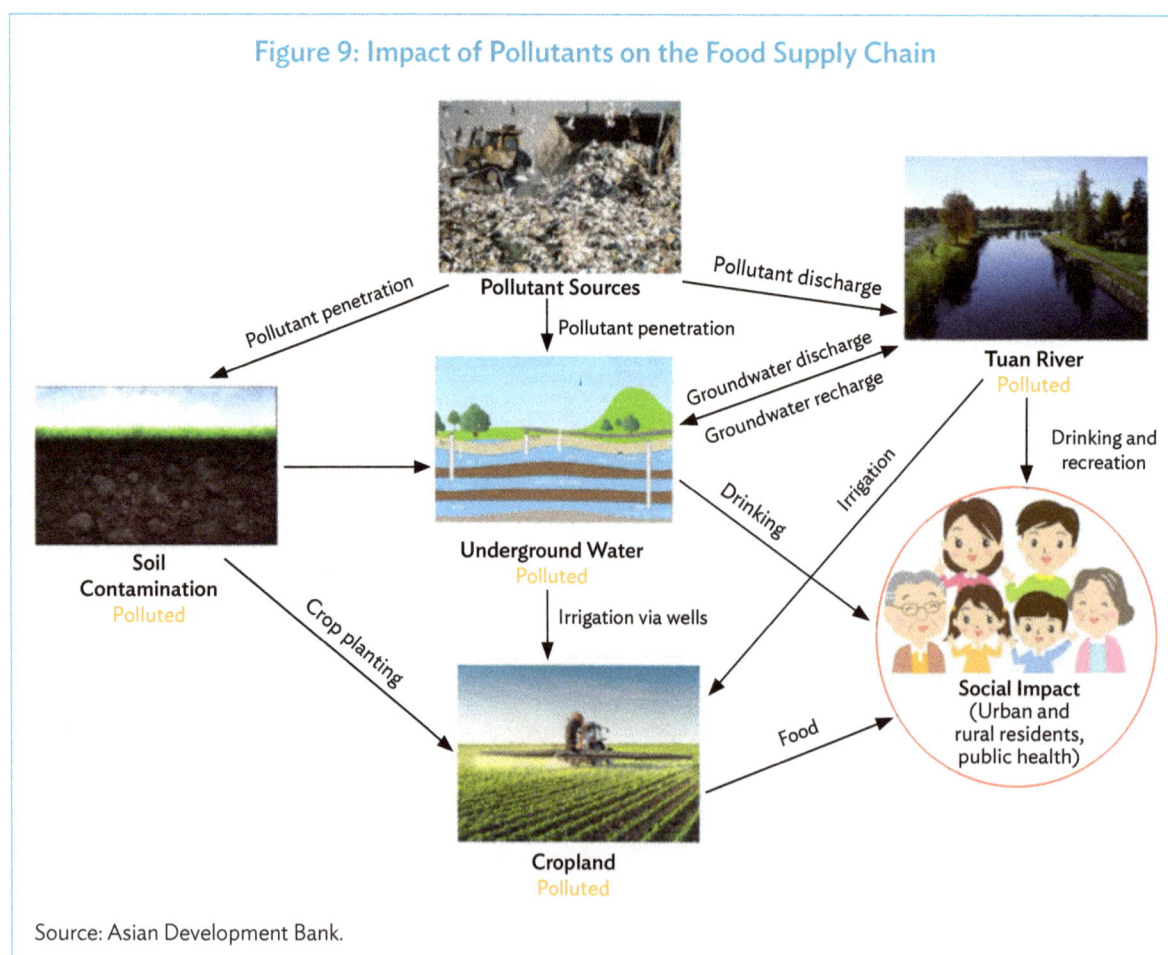

Figure 9: Impact of Pollutants on the Food Supply Chain

Source: Asian Development Bank.

---

33   A *mu* is a Chinese unit of land measurement, which is equivalent to 1/15 ha or 666.67 square meters.

66.	Statistical data for 2018 indicate that vegetables were grown on 517,500 *mu* (34,500 ha) of land within Dengzhou City, with total production of approximately 1.26 million tons, much of which was distributed and sold to urban residents in Dengzhou and elsewhere.

67.	Suspended solids discharged from sewers, storm pipes, and NPSs have accumulated in sediments underlying the Tuan River and have severely degraded the health of aquatic habitat. For example, sediment deposition transforms the habitat of benthic plants and animals, and suspended solids block sunlight and depress the growth of beneficial aquatic vegetation. The Tuan River is almost ecologically dead, which impacts some people's livelihoods, especially fishers who still rely on fishing its waters. Excessive siltation also impairs drinking water treatment processes and quality of the supplied water.

## F.	Taking Action: Addressing the Urban–Rural Water Issues in Dengzhou City

68.	Recognizing that ongoing urban development and issues on pollution and waste management seriously threaten living conditions, quality of life, sustainable development, and the environment, the DCG has embarked on an extensive program to improve conditions in the Tuan River and improve the livelihoods of urban and rural residents. The Asian Development Bank (ADB) is supporting the DCG's program, following the URWLs approach to water and river management that has kicked off and provided guidance to the implementation of the rural vitalization plan of the DCG, and has introduced the new approach to the ADB–PRC Yangtze River Economic Belt (YREB) strategic framework with different innovative components.[34] ADB's assistance, through the Henan Dengzhou Integrated River Restoration and Ecological Protection Project, is described in Chapter IV.

34	ADB and the Government of the PRC agreed to adopt a framework approach to strategically program ADB's lending support for development initiatives in the YREB, with priority given to (i) institutional strengthening and policy reform; (ii) ecosystem restoration, environmental protection, and management of water resources; (iii) inclusive green industrial transformation; and (iv) construction of an integrated multimodal transport corridor. ADB. 2018. *Framework for the ADB's Assistance for the YREB Initiative, 2018–2020.* Manila.

III

Government Strategies, Policies, and Initiatives

69.   All developing countries experience differential rates of development in urban and rural areas, which lead to particular social and environmental issues that governments must contend with. In developing countries, socioeconomic progress is commonly driven by development in the urban areas, and the rural areas are usually left behind. This experience is certainly not unique to the People's Republic of China (PRC). Because of the dramatic rate of socioeconomic progress in the PRC, the gap between urban and rural living conditions has grown rapidly, and severe environmental problems have arisen. The PRC is, therefore, a good model to use to examine river pollution, an issue relevant to all developing countries. It is instructive to review the national strategies, government policies, and initiatives being implemented in the PRC to manage the problems associated with a widening urban–rural divide in order to restore social equity, conserve the environment, and promote sustainable national and regional development. In this report, the emphasis is on urban–rural linkages in water management as an approach to restore river health and rehabilitate the river environment. But water management issues cannot be considered in isolation from other issues associated with management of nonwater-related urban–rural dependencies.

A.   Socioeconomic Urban–Rural Challenges in the People's Republic of China

70.   Sharing the prosperity of socioeconomic development more equitably between urban and rural communities has become an emerging issue for social harmony in the PRC. It is first necessary to identify urban–rural linkages and challenges if those challenges are to be effectively addressed.

1.   Widening Urban–Rural Gap

71.   The first challenge is poverty in the rural areas. The remaining 30 million citizens of the PRC living in poverty are mainly in the countryside.[35] Rural poverty alleviation is a precondition for rural rejuvenation and narrowing of the urban–rural gap. A major reason for rural poverty is that the agriculture sector is poorly structured and in need of modernization. Outcomes are poor product quality and concentration on primary production only. This, together with decreasing prices for agricultural products and limited access to market information, makes it very difficult for farmers to maintain their livelihoods.

72.   The second urban–rural challenge is the promotion of green development. To guarantee food safety and control air, water, and soil pollution, the living conditions of rural residents need to be improved, agricultural practices need to change, and the sources of pollution need to be better managed. Rural areas face serious ecological and environmental pollution problems because of the slow development or lack of progress in rural management practices, excessive use of pesticides and chemical fertilizers, and inappropriate treatment of livestock and poultry wastes.[36] In rural areas, environmental awareness is weak and priority is given to the pursuit of economic returns, which will not be sustainable unless the natural resources (including water

---

35   In the PRC, poverty is defined as income less than $1.90 per day.
36   From 2000 to 2014, the total value of agricultural production more than tripled, food production increased by 30%, and the consumption of chemical fertilizers increased by more than 40%.

resources) are conserved. Waste management is generally inadequate, as facilities like wastewater treatment plants (WWTPs) and solid waste landfills are not well developed. By the end of 2016, the treatment rate of wastewater in the PRC was 92% in urban areas and only 22% in rural areas, demonstrating a large urban–rural gap between waste management and waste treatment capacity.

73.   One of the most challenging tasks is to strengthen rural governance, particularly in environmental management. Because of the impact of the booming market economy and the accompanying massive influx of population into cities, many rural areas have been neglected and depopulated. Administrative skills within local governments must be adequate to perform tasks such as the following: support effective management and innovation, collect relevant data and information, identify and promote economic opportunities, and arbitrate disputes among villagers. However, administrative skills and management methods among government staff are very often more backward in rural areas in comparison to urban areas.

74.   Another important challenge is that public infrastructure remains inadequate in rural areas. Lack of modern water supply facilities risks drinking water safety, and waterborne infectious diseases may spread among rural residents. Outdated power supply infrastructure causes poor service performance in rural areas, with inefficient facilities resulting in high energy consumption, service interruptions, and poor power quality (particularly low voltage caused by old power grids). Despite the inferior service, the average price of electricity is generally higher in rural areas. This impedes the use of electricity in the production of rural goods and services, and increases living costs for rural residents. Unpaved roads in poor condition constitute another rural infrastructure issue, especially for townships and villages in mountainous areas. In general, constructed roads in rural areas are poorly maintained because of inadequate engineering management and quality control.

75.   The low level of education in rural areas is another major challenge. Despite outstanding national achievements in education—with adult literacy rising from 65% to 96% since the 1980s, and more than 60% of high school graduates now attending university (compared to 20% in the 1980s)—a significant gap between urban and rural education remains. More than 70% of urban students are admitted to college, compared to less than 5% of rural students. One reason is that the average income of urban residents is three times that of rural residents. Rural children may be needed for labor on family farms. Many more capable and experienced teachers also find it less attractive to work in rural areas. Moreover, as more than 250 million people have migrated from rural to urban areas in search of higher paying jobs since the 1980s, more than 60 million rural children of school age have been left behind.

76.   There is a clear gap in the standard of social security for urban and rural communities in the PRC. This applies to social support measures such as pensions and medical insurance, which do not meet the needs of farmers and rural residents or match the standard of social support available in urban centers.

77.   Finally, rural communities have more limited access to finance. Capital investment by the Government of the PRC in the rural social security system is comparatively low. Traditional capital sources of rural social security have mainly depended on subsidies for families in need provided by the rural collective. Since the household contract responsibility system was introduced in the 1980s, which allows families to keep a proportion of their farm produce for private income, sources of income for most of the rural collective economic organizations have been cut, and the availability of communal funds for the collective subsidy has declined. In relatively poor communities, local governments have inadequate capacity to make up the capital deficit for social security needs.

2.   Rural Vitalization Plan

78.   One of the responses by the Government of the PRC to bridging these urban and rural gaps was a plan to vitalize the country's rural areas. A rural vitalization strategy was first proposed in 2017 and formally became one of the national strategies after the State Council issued the National Strategic Plan for Rural Vitalization, 2018–2022 in September 2018.[37] Compared with previous strategic plans that favored urbanization and caused an imbalance between urban and rural areas in terms of development focus, the rural vitalization plan gives priority to formulating programs for rural regions tailored to redressing the issues arising from the past urban–rural imbalance. Rural regions must function as areas that meet the growing agricultural production needs of the nation, and to do that successfully and sustainably requires social improvements, cultural preservation, and environmental and ecological conservation of natural resources. Recognizing the interdependency of urban and rural areas, the rural vitalization plan aims to vitalize vast rural regions in the country so they are able to play a more effective role in supporting urban populations and advancing the national economy. One important aspect of rural vitalization is that public services in rural regions need to be improved to narrow the gap between urban and rural areas. However, to be a successful role model, the rural vitalization plan needs to be integrated with urban development rather than act as a stand-alone approach.

B.   Water Resources Management Challenges and Issues in the People's Republic of China

79.   In 2018, ADB published a country water assessment for the PRC, which gives an in-depth review of water resources management issues and governing laws, regulations, policies, and strategies in the PRC (footnote 13). Reference is likewise made to the analytical framework for measuring water security developed by ADB and the Asia-Pacific Water Forum as part of the Asian Water Development Outlook series.[38]

C.   Water and Natural Resources Management: Laws, Policies, and Strategies

80.   The water resources issues referred to in paras. 11–13 have prompted the Government of the PRC to promulgate numerous laws and regulations, and initiate many policies and programs to address these issues. The most important in the context of urban–rural water linkages (URWLs) are described in paras. 81–87.

1.   Legislation and Policy

81.   The foundation of water resources management is the Water Law of the PRC, which took effect in October 2002. It stipulates goals, guidelines, principles, systems, and procedures for water resources management in the country. The Water Law is based on the principles of integrated water resources management (IWRM).

---

[37]   Government of the PRC, State Council. 2018. China Releases Five-Year Plan on Rural Vitalization Strategy. News release. 26 September; and Government of the PRC. 2018. *National Strategic Plan for Rural Vitalization, 2018–2022*. Beijing.

[38]   ADB. 2007. *Asian Water Development Outlook 2007: Achieving Water Security for Asia*. Manila; ADB. 2013. *Asian Water Development Outlook 2013: Measuring Water Security in Asia and the Pacific*. Manila; and ADB. 2016. *Asian Water Development Outlook 2016: Strengthening Water Security in Asia and the Pacific*. Manila.

82.   In January 2011, the State Council issued the Opinion on Implementing the Most Stringent Water Management System. This was the most important guidance provided by the State Council in relation to water resources management since the promulgation of the Water Law, with far-reaching significance for management of water resources and environmental problems in the PRC. The main content of the document covers the establishment of the "three red lines" policy and the enforcement of this policy through the detailed "four systems" plan.[39] Another red line was added by the ecological redline policy, which aims to rigorously control urbanization and industrial development in specified zones requiring special protection.[40]

83.   Since 2000, market-based instruments such as pollution rights trading and eco-compensation have attracted growing government attention. This demonstrates that the PRC's reform toward a market economy has expanded into fields of environmental and natural resources management.

84.   The Government of the PRC has been promoting eco-compensation schemes nationwide. Eco-compensation involves the use of market-based environmental policy tools and includes what is often referred to as payment for environmental services. In the PRC, eco-compensation also encompasses fiscal transfer schemes between provinces to share funding, responsibilities, and tasks on environmental management—especially if the transfer of ecological services crosses administrative boundaries, including services that conserve water, protect water quality, or prevent soil erosion within large river basins.[41] Essentially, eco-compensation mechanisms seek to compensate those who sustainably manage the natural resources in exchange for their provision of one or more ecosystem services—e.g., water filtration, erosion control by maintaining forest cover, and improved watershed land use and management practices.[42]

2.   Water Pollution Action Plan and River Chief System

85.   The Water Pollution Action Plan was issued by the PRC State Council in April 2015 to address policy directions. Its primary objective is the environmental protection of major rivers and tributaries through improvement of water quality and tightened control of pollution sources. The plan takes into consideration existing landscapes and the ecological protection and water management requirements to prevent water pollution. The goals for 2020 require significant improvement in the water quality of rivers and lakes nationally to increase available drinking water, decrease the number of incidents related to water pollution, mitigate the impacts associated with groundwater overexploitation, and significantly improve environmental conditions.

86.   The Water Pollution Action Plan includes the following components: (i) regulation of pollutant emissions, (ii) promotion of economic restructuring and upgrading to reduce emission of pollutants, (iii) water resources conservation, (iv) strengthened support for science and technology, (v) promotion of market mechanisms, (vi) compliance with and enforcement of environmental laws, (vii) strengthened management of water and natural resources, (viii) enhanced safety of aquatic environments, (ix) implementation of clearly defined management roles and responsibilities, and (x) increased public participation and social engagement.

---

39   The "three red lines" policy sets compulsory targets for (i) controlling total water use, (ii) improving water use efficiency, and (iii) capping total water pollution discharge loads (footnote 13, pp. 47–48). The "four systems" refer to (i) a system to assess the available water resources and determine the quantitative target for total water use; (ii) a system to determine the water use efficiency red line based on effective management of water demand, but consistent with meeting the future water needs for economic and social development; (iii) a system for restricting pollution in defined water function areas (river reaches, lakes, and reservoirs); and (iv) a system for accountability and assessment.

40   The ecological redline policy was incorporated into the PRC's Environmental Protection Law in 2014. Its overall target is the protection of the integrity of essential ecosystems to secure the delivery of diverse yet interconnected ecosystem services required to meet the needs of various stakeholders in the ecological redline areas. Y. Bai et al. 2016. New Ecological Redline Policy (ERP) to Secure Ecosystem Services in China. *Land Use Policy*. 55. pp. 348–351.

41   Q. Zhang and M. T. Bennett. 2011. *Eco-Compensation for Watershed Services in the People's Republic of China*. Manila: ADB.

42   ADB. 2016. *Addressing Water Security in the People's Republic of China: The 13th Five-Year Plan (2016–2020) and Beyond*. Manila.

87.   In 2007, the Government of the PRC assigned local government officials to be "river chiefs" to address pollution woes of a blue-green algae outbreak in Taihu Lake, Jiangsu Province. The basic concept was (i) to incentivize officials to focus attention on water resources management within their districts to achieve a better balance between economic and environmental policies, and (ii) to make officials accountable to meet water quality and related environmental targets. The concept was later adopted in other regions of the PRC to improve water governance, strengthen enforcement of environmental policies, and enhance coordination of different government agencies. Based on the experience gained in those pilot schemes, in 2018, the central government established the river chief system nationwide, and this new measure has been implemented across most provinces. Around the country, more than 300,000 officials have been appointed as river chiefs working at the provincial, city, county, and township levels; another 760,000 have been designated at the village level.

D.   Yangtze River Economic Belt Development Program

88.   The Yangtze River Economic Belt (YREB) subregion comprises nine provinces and two specially administered cities. It includes three major urban agglomeration areas: (i) the Yangtze River Delta Global Megacity Agglomeration, focused on Shanghai; (ii) the Middle Yangtze City Cluster, focused on Wuhan; and (iii) the Yangtze Upper Reaches, focused on Chongqing and Chengdu. Because of the YREB's geographic and demographic advantages, the government has identified it as a key engine of growth in pursuit of the PRC's economic development.

89.   The Yangtze River Basin has experienced rapid development since the 1990s, particularly in the delta area around Shanghai. Economic growth in the middle and upper reaches of the basin has been slower and below potential. These reaches continue to deal with substantial development challenges arising from (i) slow progress toward green development and economic diversification; (ii) inadequate integration of ports, waterways, and intermodal logistics; (iii) increasing pollution and pressure on natural resources; and (iv) weak institutional coordination for strategic planning. The Yangtze River Basin faces a growing imbalance between economic development and environmental protection.

90.   To address these challenges, the national government formulated the YREB Development Plan, 2016–2030, which stipulates the promotion of green development and the prioritization of ecological protection as the guiding principles for future YREB development.[43] While the plan is robust in terms of environmental conservation and economic development, it can be strengthened by incorporating urban–rural integration to maximize the benefits.

91.   ADB has agreed with the Government of the PRC to adopt a framework approach and provide approximately $2.0 billion of funding during 2018–2020 to strategically program ADB's lending support for development initiatives in the YREB. There are four priority areas: (i) environmental protection, ecosystem restoration, and water resources management; (ii) green and inclusive industrial transformation; (iii) construction of an integrated multimodal transport corridor; and (iv) institutional strengthening and policy reform.

92.   The Henan Dengzhou Integrated River Restoration and Ecological Protection Project (footnote 3) is part of ADB's contribution to the YREB Development Plan. Dengzhou City and the Tuan River are in the Yangtze River Basin. The urban–rural linkages described in this report address the priority areas of ADB in the YREB—i.e., ecosystem restoration, environmental protection, and water resources management.

---

43   Government of the PRC. 2016. *Outline of the YREB Development Plan, 2016–2030*. Beijing.

E.   Yellow River Basin Program

93.   The Government of the PRC has expressed its intention to draft a special development plan for the Yellow River Basin. The Yellow River is the second longest river in the PRC and has played a major historical role in Chinese civilization. It supports about 12% of the PRC's population, irrigates about 15% of arable land, generates 14% of national gross domestic product, and supplies water to more than 60 large cities.[44] Problems requiring urgent attention include the fragile environmental conditions in many parts of the river basin, water scarcity, degraded water quality, and water security for future socioeconomic development through improved water resources management and tighter management of industrial water demand.

94.   An integrated approach to sustainable agriculture and rural vitalization is one of the government's priorities in the Yellow River Basin, and the URWLs will play a critical role in achieving these goals. ADB is in discussions with the Government of the PRC to determine how best it can support this proposed program of regional development in the Yellow River Basin and encourage more sustainable management approaches. There are many modern concepts and means to promote more sustainable management of water resources, and one such approach is awareness of URWLs. Integrated urban–rural development is a common theme for ADB, as reflected in its Strategy 2030 (footnote 8), and the Government of the PRC.

F.   Institutional Reform

95.   There is ongoing institutional reform within the central and local governments. In early 2018, the Government of the PRC started restructuring its ministries, with some adjustments in their respective mandates. The Ministry of Natural Resources was established in April 2018 and absorbed eight former ministries and departments (including the Ministry of Land and Resources). The Ministry of Emergency Management was formed, which is mandated to deal with all water-related and other disasters. It absorbed the functions of 13 relevant departments and agencies under different ministries. The Ministry of Ecology and Environment replaced the former Ministry of Environmental Protection and consolidated some of the functions and personnel of other ministries or agencies. These new ministries and functions will take over some of the tasks of the still existing Ministry of Water Resources. Although responsibilities related to water and the environment are still fragmented among ministries, the ongoing institutional reform that began in 2018 may help establish proper coordination among national ministries and local governments.

G.   Assessing Urban–Rural Water Linkages
in Government Programs

96.   Integrated urban–rural development is a repeated theme in technical discussions and in government thinking in the PRC; yet, present laws, regulations, policies, and strategies for water resources management do not explicitly address URWLs. Despite that, all the basic conditions and characteristics for a URWLs approach are treated in various provisions. An integrated urban–rural approach to water resources management is clearly consistent with recent advances in government policies and management strategies. To mention a few: (i) the IWRM approach, which needs to be followed in the Water Law; (ii) the holistic approach to sustainable water resources management, as established in the Opinion on Implementing the Most Stringent Water Management System through the three red lines policy (para. 82; and footnote 39); (iii) priorities to reduce pollution and restore healthy water systems under the Water Pollution Action Plan; (iv) the imperative for ecological protection and conservation, as stipulated in the ecological redline policy

44   *Xinhua News Agency.* 2019. Xi Focus: Xi Stresses Ecological Protection and High-Quality Development of Yellow River. 19 September.

(para. 82; and footnote 40); and (v) improved enforcement of relevant provisions to protect rivers and lakes (the river chief system). The rural vitalization strategy is another important stimulus to the adoption of a URWLs approach, as it aims to reduce the gaps in living standards and development opportunities between urban and rural areas.

97.   Given that these existing laws, regulations, policies, and strategies already address key parts of URWLs and propose an integrated management approach, specific legislative amendments or strategies to further promote a URWLs approach in the PRC seem unnecessary. The best way to advocate URWLs is to showcase successful applications. The Tuan River example, discussed in Chapter II, and its related ADB-supported project in Dengzhou City, Henan Province (footnote 3), outlined in Chapter IV, can serve as case study in this regard.

Embedding Urban–Rural Water Linkages in Project Design

IV

98.   In response to the issues and problems noted in Chapter II, the Dengzhou City Government (DCG) embarked on a program to improve conditions in the Tuan River Basin, with the ultimate goals of sustainable economic growth in the basin area and improved livelihoods of urban and rural residents. The urban–rural water linkages (URWLs) and the river health rehabilitation of the Tuan River were central to achieving these goals and were incorporated by the Asian Development Bank (ADB) in the project design of the Henan Dengzhou Integrated River Restoration and Ecological Protection Project (footnote 3), which aims to support the DCG.

99.   The Henan Dengzhou project primarily focuses on a holistic environmental management approach, applying a mix of structural and nonstructural interventions in urban and rural areas of Dengzhou City, as a demonstration of integrated urban–rural water resources management. It showcases how a URWLs approach combined with integrated water resources management (IWRM) practices can benefit the river environment and ecological systems. The concepts applied in this project can be utilized as a model for development in other rivers of the People's Republic of China (PRC), and in other developing countries.

100.   The project addresses the issues of land use, water withdrawal, and pollution with a URWLs approach (Figure 5 and Figure 6). Three sets of project components were formulated to address these issues. A fourth set includes institutional strengthening measures necessary for successful implementation. The major project components are as follows:

(i) Land use
  (a) Tuanbei New District ecological and water improvement
  (b) Tuan River ecological restoration
  (c) Xingshan forest plantation for soil erosion control
(ii) Surface water and groundwater withdrawal
  (a) Implementation of water supply systems in rural areas
(iii) Pollution
  (a) Wastewater management projects in both urban and rural areas
  (b) Solid waste management in rural Rangdong Town
  (c) Bio-shield for nonpoint pollution and soil erosion control
(iv) Institutional capacity strengthening

A. Land Use

101.   The first set of project components addresses land use issues in the urban and rural areas of Dengzhou City. Land use is a driver for change in the river morphology and water pollution. These project components include support for erosion control, rehabilitation of the river environment, and flood mitigation. In urban areas, an approach consistent with the sponge city (footnote 10) and low-impact development (LID) concepts guides the engineering design. The basic principle of this approach is to make use of natural processes and manage runoff from rainfall at the source. This is accomplished through sequenced implementation of measures to enhance infiltration and reduce surface runoff, lessen stormwater runoff, and treat stormwater runoff to remove pollutants. Examples of project components related to land use are described in paras. 102–113.

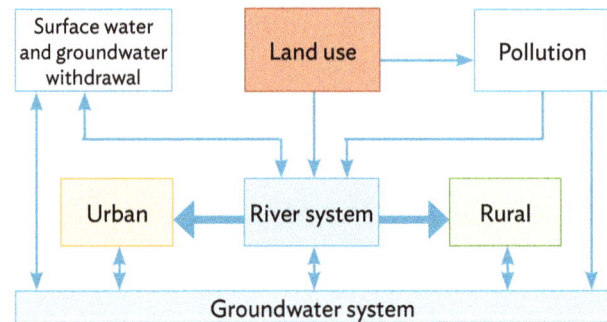

1.   Tuanbei New District Ecological and Water Improvement

102.   The Tuanbei New District is situated on the north bank of the Tuan River in Dengzhou City (opposite the main urban center) and covers an area of about 20.1 square kilometers (km^2). Most of the area consists of farmland and undeveloped land, with small areas of low-density residential dwellings and rural villages. During storm events, runoff from the farmland transports pollutants (such as sediments, nutrients, oil, metals, hydrocarbons, and other toxic substances) before entering ditches or storm drains, which discharge into the Tuan River or its tributaries. Applying the sponge city or LID approach will reduce and attenuate the stormwater runoff, thereby lessening the risks of flooding. It will also reduce the pollutant load into local receiving waters, improving water quality in the Tuan River. As groundwater in the project area has been overexploited, having more permeable ground surfaces will increase infiltration and play a positive role in recharging groundwater aquifers.

103.   **Tuanbei New District Green Corridor Park.** The proposed Tuanbei New District Green Corridor Park will include a number of water management and landscaping works, such as the Weiming Channel (a landscaped waterway and wetland), tree planting, sunken greenbelt rain gardens, and cultural leisure facilities. The total park area is 65.3 hectares (ha). The park layout (Figure 10 and Figure 11) includes an ecological conservation zone, a shady grassland, a recreational leisure zone, a cultural park, sports facilities, a natural children's park, a riverside ecological zone, and a floating plank walkway. The development will create a transitional buffer zone or green corridor to provide leisure and recreational opportunities for the urban and rural populations, and will intercept and treat pollution transported toward the river.

104.   Ecological revetments will be constructed beside the open channels, water diversion canals, and the Weiming Channel to stabilize banks and reduce bank erosion. Vegetation zones will be installed beside the Weiming Channel within the park. The design length of the Weiming Channel is 2,940 meters. At its southern end, the channel overflows to the Tuan River. The water surface area of the channel will be 9.8 ha, with total water volume of 103,000 cubic meters (m^3).

105.   **Tuan River North Shore Green Park.** Another proposed project, the Tuan River North Shore Green Park, will cover an area of 39.5 ha. The park design includes three major zones: (i) an LID demonstration zone in the west, (ii) a leisure or entertainment zone in the middle, and (iii) a constructed wetland education zone

Figure 10: Layout of the Proposed Tuanbei New District Green Corridor Park

Source: Dengzhou City Government. 2019. *Initial Environmental Examination (Draft): Henan Dengzhou Integrated River Restoration and Ecological Protection Project in the People's Republic of China* (prepared for the Asian Development Bank). Adapted from Figure V-39, p. 143.

Figure 11: Internal Landscape Design of the Proposed Tuanbei New District Green Corridor Park

Source: Asian Development Bank. People's Republic of China: Henan Dengzhou Integrated River Restoration and Ecological Protection Project.

in the east. This project intervention in the Tuanbei New District will reduce runoff and effluent discharges to the river by an estimated 12.6 million cubic meters (mcm) per annum, and substantially reduce pollution of the underlying groundwater aquifer.

106.   **Drainage improvement.** Residents of the Tuanbei New District are vulnerable to waterlogging or local flooding. The stormwater drainage network is incomplete and the capacity of former creeks and ditches has progressively diminished because of land use encroachment (for farmland or during urbanization), sediment deposition, and dumping of rubbish. The project will implement a drainage system to collect stormwater by a pipe network of drains, a water diversion channel, and two main drains. This will reduce the risk of flooding and improve public safety. The DCG will demolish a small dam it had constructed to pool water in a fruitless attempt to improve the appearance of the degraded urban reach of the Tuan River. A revised river improvement plan is under preparation by the DCG.

107.   Environmental improvements to the river and urban waterfront, the Tuanbei New District Green Corridor Park, and the Tuan River North Shore Green Park will create more attractive living conditions and, based on experience elsewhere, raise values of adjoining property. Revenue from land sales will partly offset expenditures on these measures, which are mainly for operation and maintenance of the established infrastructure.

108.   **Stormwater quality improvement.** Implementation of LID practices in urban areas of Dengzhou City during the project will reduce polluted stormwater reaching the Tuan River and improve water quality of the lower Tuan River by reducing sediments, particulate organics, and contaminants in stormwater runoff before entering the river or its tributaries. Better water quality will increase property values and lower government cleanup costs. The LID measures are expected to limit average concentrations in stormwater runoff of chemical oxygen demand to 20.0 milligrams (mg) per liter (L) or less, ammonia nitrogen (NH_3-N) to 1.0 mg per L or less, and total phosphorus to 0.2 mg per L or less.

2.   Tuan River Ecological Restoration

109.   **Dredging.** A 13.8-kilometer (km) reach of the Tuan River downstream from Dengzhou City contains bed and bank deposits that are highly polluted from sediment deposition and accumulation of urban construction wastes. Sediment sampling and laboratory testing indicated low concentrations of heavy metals and pesticides in the bed deposits, but high concentrations of organics, nitrogen, and phosphorus, which derive mainly from domestic sewage discharge and nonpoint source (NPS) runoff from agricultural activities. Dredging to an average depth of 0.3 meters is proposed at discrete locations along this 13.8 km river reach—with a combined length of 3.8 km—to remove about 207,000 m³ of polluted river sediments. This is an environmental protection measure to improve river water quality by removing contaminated sediment deposits on the riverbed. It is projected to improve the river water quality by reducing NH_3-N by 12.9%, reducing biochemical oxygen demand (BOD) by 8.1%, and increasing dissolved oxygen of the water column by 2.3%. Discarded construction materials and old earth bunds will also be removed along this reach of the river.

110.   In addition to its effects on water quality improvement, the bed sediment dredging and excavations will enlarge river channel cross-sections, which will increase the flow conveyance capacity of the river. Based on hydraulic modeling analysis, this project component will increase the Tuan River flood conveyance capacity by 5%–10%.

111.   **Riverbank protection and repair.** The project will also raise crest levels of selected levee embankments to meet the required 1-in-20 year flood protection standard of the Tuan River. Levee construction will use green measures for ecological slope protection. Downstream of Dengzhou City, most of the riverbanks are modified or cleared by human activities. Riverbanks have become highly unstable and prone to landslips or undercutting erosion by river flows. The project will strengthen riverbanks with ecological embankments. These include stone-filled gabions for riverbank protection at seven separate locations, with a combined length of 3.7 km. The levee works will protect an estimated 1,361 households and 16,166 *mu* (1,078 ha) of farmland, with significant reduction in flood damages.

112.   **Vegetation of riparian zones.** Planting of aquatic plants in two artificial wetlands and at the confluences of the Tuan River with the Huangqu and Zhao river tributaries will treat polluted runoff. The aquatic plants absorb inorganic nutrients, such as nitrogen and phosphorus, and capture toxic and harmful substances, including heavy metals. They also produce oxygen through photosynthesis, increasing dissolved oxygen concentrations in water and improving the capacity of the river to assimilate pollutants. Figure 12 illustrates some of the proposed river improvements. The nature-based solution to improve the quality of the river environment will also improve the riverfront design. Revenue from real estate will partly offset expenditures on these measures, mainly for operation and maintenance of the established infrastructure.

Figure 12: Design of Tuan River Improvement Works

Riverbank Improvement

Aquatic Plants in River

Riverbank Improvement

Riverbank Protection

Source: Asian Development Bank. People's Republic of China: Henan Dengzhou Integrated River Restoration and Ecological Protection Project.

3.   Xingshan Forest Plantation

113.   The Xingshan forest plantation project is located near the head of the canal of the middle route of the South-to-North Water Diversion Project (SNWDP) that diverts water from Danjiangkou Reservoir.[45] The project will be beneficial for soil erosion control, carbon capture, and management of water quality of the reservoir. The proposed area of the tree plantation is 800 *mu* (53.3 ha). It will help reduce erosion problems in the Xingshan Hill area, where soil is eroding from farmlands and other land with poor vegetation cover. Ancillary works include an irrigation system for the plantation and a service road for forest fire protection. Trees absorb carbon dioxide and are vital carbon sinks that help tackle climate change.[46] Afforestation in the Xingshan project will contribute to the commitment of the Government of the PRC on carbon emission reduction in fulfillment of the Paris Agreement, which the government signed in 2015, and is aligned with the PRC's Intended Nationally Determined Contributions to reduce its carbon dioxide emissions after peaking by 2030.

B.   Surface Water and Groundwater Withdrawal

114.   The second set of project components concerns withdrawals of surface water and groundwater. The project mainly addresses water supply in the rural areas of Dengzhou City. Because of lack of clean water sources and reliable water supply, rural areas rely on highly polluted groundwater for drinking purposes. There is an urgent need for water treatment plants and affiliated piped water supply networks to provide safe drinking water to the rural residents.

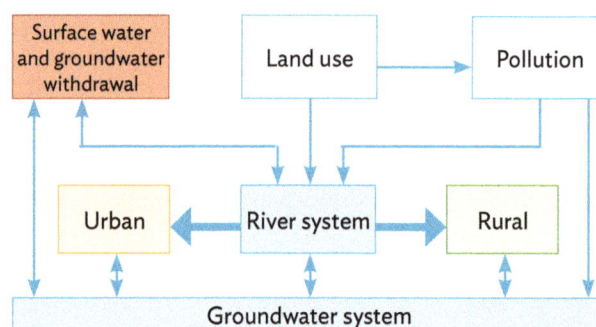

115.   Dengzhou City has poorer water supply than the national and provincial norms. The share of the population serviced by a community water supply system, the coverage of connections within water supply areas, and the quality of drinking water are all well below the Henan Province and national averages. Rural potable water in Dengzhou City is largely unsafe.

116.   Two township water supply plants (WSPs) proposed at Sangzhuang and Jiulong will service township and/or rural areas of Dengzhou City. These WSPs will supply 60,000 m³ per day of drinking water to 10 townships of Dengzhou City to improve the security and sustainability of rural water supply in rural towns and villages. Both WSPs will draw water from the south–north water transfer canal, where the standard of water quality is already class II, i.e., suitable for domestic use (footnote 14). The construction of the WSPs will significantly reduce the overexploitation of groundwater in rural areas and avoid reliance on polluted groundwater as the source of water supply.

117.   The Sangzhuang WSP will have a capacity of 30,000 m³ per day, servicing residents in five rural towns. Service facilities include a 305 km piped water distribution network, a booster pump station, a chlorination station, and a 750-meter diversion pipeline from the water source. The Jiulong WSP will also have a capacity of

---

45   The reservoir, constructed in the 1950s, has a surface area of 1,050 km², storage capacity of 29.05 cubic kilometers, and a watershed area of 95,200 km². Besides irrigation, flood control, and hydropower generation, it also supplies water through a 1,274 km stretch of canal to 14 cities in the northern PRC, benefiting more than 50 million people in the water-scarce cities of Beijing and Tianjin, and the provinces of Hebei and Henan under the SNWDP.

46   Each year, a large tree inhales about 20.3 kilograms of carbon dioxide and exhales enough oxygen for a family of four.

30,000 m³ per day, servicing another five rural towns. Service facilities include a 310 km piped water distribution network, three booster pump stations, and a 1.2 km diversion pipeline from the water source.

C.   Pollution

118.   The third set of project components is aimed at sources of river and groundwater pollution. Project activities include construction of wastewater treatment plants (WWTPs), measures for managing solid waste, and measures to control pollution from agriculture.

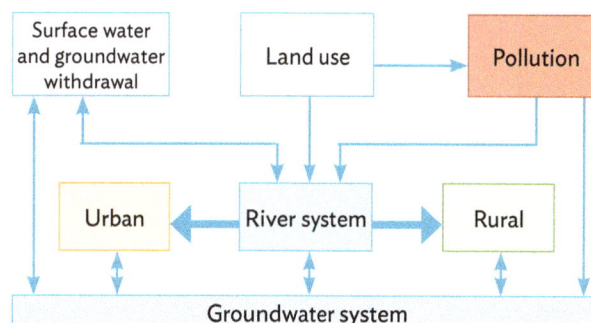

1.   Wastewater Treatment Plants

119.   The Tuanbei New District of Dengzhou City has a population of 20,000 but has no WWTP yet. It is expected that population will grow to 180,000 by 2030 and to 200,000 by 2040. Construction of three WWTPs is proposed under the project, which will be adequate to cope with wastewater treatment needs of the rapidly urbanizing areas as well as adjoining rural areas. The proposed Tuanbei WWTP has a treatment capacity of 30,000 m³ per day, capable of servicing approximately 180,000 residents of the Tuanbei New District. Another two rural WWTPs are proposed: (i) in Rangdong Town, with a treatment capacity of 2,000 m³ per day; and (ii) in Jitan Town, with a treatment capacity of 1,500 m³ per day.

120.   Wastewater management in both urban and rural areas will reduce sewage pollutant loadings into the Tuan River, decreasing the accumulation of nutrients, toxic substances, and heavy metal in the river sediments and helping to restore river health. The proposed WWTPs are expected to make significant improvement to water quality in the lower Tuan River by reducing total phosphorus by 20%, total nitrogen by 19.7%, NH_3-N by 37%, and organic contaminants (BOD) by 16%. In the national context, construction of the WWTPs supports the PRC's 13th Five-Year Plan, 2016–2020, which has a target of enhancing the wastewater treatment facilities and wastewater collection systems to mitigate harmful effects from wastewater.

2.   Solid Waste Management in Rangdong Town

121.   The solid waste management (SWM) project includes treatment of vegetable and fruit wastes generated by a farmer's market using a pilot biological treatment technology. Other sources of organic wastes to be treated include domestic leftovers; expired food; and wastes from places like restaurants, hotels, enterprise canteens, and shops where food is processed. Organic waste output from the township is approximately 20 tons per day.

122.   The project will apply Bio-Star microbiological degradation technology (produced by Les Traitements Bio-Bac Inc., a company specializing in biological treatment of organic waste) for treating leftovers and perishable organic wastes. This advanced treatment method has advantages, including high maturity, low energy consumption, low cost, high degradation efficiency, minimum secondary pollution, and reliable and user-friendly operation. Liquid effluent from the bioreactor will be further treated in the Rangdong WWTP, with proper disposal of residual sludge.

Farmer's market in Rangdong Town. Vegetable and fruit wastes from the farmer's market are being treated under a pilot waste management subproject (photos from the Dengzhou City Government).

3.   Bio-shield for Nonpoint Source Pollution Control

123.   Most of the Tuan River Basin is farmland or cultivated land, which is the source of NPS pollution. Both domestic and international experiences have proven that riparian vegetation, buffer zones, or bio-shields provide practical green technology regarded as a best management practice for managing agricultural NPS pollution. Under the project, a green buffer belt will be implemented along the banks of the lower Tuan River. The riparian buffer zone will grow various trees, shrubs, and grasses on both sides of the riverbank to restore diverse riparian vegetation communities. Its function is to reduce and filter pollutants from farmland NPSs that drain directly to the river.

124.   The bio-shield technique minimizes topsoil erosion and nutrient loss. Pervious buffer belts reduce the volume of runoff to surface water bodies and enhance groundwater recharge. They also trap sediments and provide opportunities for water quality improvement by breaking down contaminants through biological activities of microorganisms in the porous media. In the long run, accumulated sediment along the bio-shield will raise the riverbank and enhance flood mitigation. The wet environment along the buffer zone increases denitrification, thus reducing groundwater contamination with nitrate. The fine sediments deposited in the riparian buffer also reduce phosphorus leaching because of fixation of phosphorus to the soil particles. Therefore, the riparian vegetated buffer zones (belts) on both sides of the Tuan River are expected to reduce total phosphorus, total nitrogen, and other pollutants and nutrients entering the river. Native plants are encouraged, as such layers provide multipurpose uses such as fodders and riverside gardening (with foot trails as appropriate). In addition, a total of 125 ha of aquatic vegetation will be planted in shallow water zones (artificial wetlands) in the lower reaches to improve the river ecology.

D.   Institutional Strengthening

125.   The project components discussed in paras. 126–132 will strengthen institutional capacity to manage and build on project achievements and sustain project benefits.

126.   **Integrating urban and rural water services management.** The large gap in the standard of services (water supply and sewerage) between urban and rural areas in the PRC stems from institutional issues. Although they still rely on government subsidies, service providers in urban areas operate on a more economically sustainable basis, charging fees to consumers that generally cover most, if not all, operational costs. Tiered tariffs are often used to recover more costs from larger consumers, and urban water tariffs have been progressively increased since 2000. In rural areas, there are no charges for water. Historically, this was because of widespread poverty in the countryside. Although there were valid social reasons for not charging fees, it has led to inferior

standards and has constrained social and economic advancement in the rural areas. Most rural water supply systems are very small-scale—either for individual households or for small communities—with infrastructure and facilities provided by local governments, but with very little maintenance or repair because of the lack of revenue. Consequently, rural water supply is unreliable. The worst pollution is in private wells sunk into contaminated aquifers; hence, poor rural public health is an associated outcome.

127.   The project conducted a willingness-to-pay survey to determine if rural residents could afford to pay a tariff and if they are willing to do so. The survey found that they are keen to receive improved water supply service delivery and are generally willing to pay a tariff comparable to the lowest-tier tariff charged for urban services. In Dengzhou City, the project is implementing two new rural WSPs in Sangzhuang and Jiulong (paras. 116–117). These two new WSPs will be operated and maintained by the same entity that provides urban water supply service for Dengzhou City, initiating an institutional and functional integration of urban and rural water services and administration. This integration was the best of several options considered by the project to provide more reliable service and a more sustainable model for future funding, operation, and maintenance of the two new rural WSPs. Integration will narrow the gaps in standard of service as well as water sector policy and administration between the urban and rural areas in Dengzhou City.

128.   **Environmental research and education center.** Management of natural resources and the environment should depend not only on the government, but should also involve public participation to strengthen it and make it more sustainable. With limited access to knowledge and information about the Tuan River Basin, public environmental awareness among Dengzhou residents is low. Public awareness should be raised by a campaign to disseminate information, conduct training, and guide initiatives to reduce pollution and protect the natural environment, including the Tuan River and its tributaries. The project is, therefore, establishing an environmental research and education center in Dengzhou City to regularly monitor the river health quality; conduct research on embedded urban–rural water issues; help adjust or upgrade the functionality of measures; and provide a means to raise awareness on the environment and the need for water resources protection to farmers, local residents, students, researchers, and government officials. The center will leverage the research capacities of academic institutions in the province and beyond and provide space (field laboratories) for researchers to learn and contribute to environmental solutions. Beyond the time frame of the project, the center will be a long-term, sustainable mechanism for raising public awareness and knowledge of the importance of natural resources in sustaining living standards. The center will also implement a system for monitoring and evaluating river health in the Tuan River Basin on behalf of the Dengzhou Environmental Protection Bureau and will support the application of river chief provisions.

129.   Five river health monitoring stations will be installed in the lower Tuan River downstream of Dengzhou's urban area. The real-time water quality monitoring will support implementation of the river chief system in Dengzhou City, improvement of water quality, management of floods, and other improvements to benefit the people of Dengzhou City. The river health monitoring system will have the following functions: automatic processing and transmission of data; automatic alarm when monitoring data (quality and quantity) is abnormal; remote diagnosis and decision support; automatic sample retention, in the case of samples exceeding a standard; and automatic quality control testing. Data will be collected automatically and transferred to a data management system to be maintained by the Dengzhou Environmental Protection Bureau.

130.   **Asset management.** Lack of asset management has substantially increased the risk of poor public services and unforeseen costs in development investments in the DCG. The project is establishing an asset management and decision support system to remedy this. Asset management involves achieving the least cost and least risk of owning and operating assets over their life cycle while meeting service standards for customers. A data center under the DCG is already equipped with advanced technology used to monitor traffic flow and power supply on a real-time basis and, thus, will run an asset management system. In a pilot intervention, this asset management system will be expanded to provide advanced asset management of water

infrastructure developed under the project. The DCG data center has trained technical and administrative staff and reports directly to the mayor's office, and has authority to coordinate with other line agencies. The data center can also help monitor river health through the real-time data received from water quality and quantity observation stations being installed along the river by the project. For security purposes, closed-circuit television cameras will also be installed at critical locations along the river corridor to detect any unauthorized activity. Asset management is a collective responsibility of multiple line agencies; therefore, an asset management coordination committee needs to be formed involving representatives from relevant agencies, which will also support the decision-making process.

131.   The key elements of the proposed asset management system (Box 3) include (i) a digital asset inventory on a geographic information system platform, with a database for historical record of services rendered for each asset's operation; (ii) an interactive real-time service delivery system that helps individual customers to call for services from their mobile devices, with a location identification function; (iii) an automatic water leakage detection system; (iv) a decision support system for asset acquisition and disposal connected to the digital asset inventory, with cost–benefit evaluation linked to the operations database; and (v) guidelines and trainings, including training for asset performance and risk assessment. This asset management system will be supported by cloud-based computing.[47] Features of the asset management system are shown schematically in Figure 13.

132.   Other capacity building within the scope of the project includes (i) strengthening community-based water management, including community-based SWM (garbage disposal); (ii) piloting manure use in crop production to reduce the use of chemical fertilizer and reduce nutrient runoff from livestock farms; (iii) promoting community entrepreneurship, mainly a community nursery and agroforestry enterprise; (iv) developing dialogue between the DCG and the SNWDP beneficiary cities to establish a fairer system for eco-compensation and accountability mechanism; and (v) drafting a water utilization plan for the city.

E.   Assessing Project Impacts

133.   Using numerical modeling, the ADB project team, in close coordination with the DCG, provided preliminary estimates about the impacts of the project in terms of pollution reduction in the Tuan River and in groundwater. The desired ultimate outcome—i.e., improving the livelihood and welfare of the resident population—cannot yet be estimated.

1.   Analytical Support to Assess the Impacts

134.   In formulating the project, the ADB project team conducted technical analysis to assess the potential effects of the project activities on the river system. This analysis relied on the development of (i) computer models that describe the hydrology and erosion processes of the Tuan River Basin; (ii) hydraulic and water quality models that describe flooding, sedimentation, and pollution in the river system; and (iii) a groundwater model.

---

47   In a cloud-based asset management system, data are stored in an online server, and applications for computations or analysis are also run from the server without needing software installed on local computers. Cloud-based asset management has wider coverage and, therefore, urban–rural integration can be facilitated. This may require procuring services from vendors (mostly from the private sector). Alibaba Group, which signed a partnership agreement with ADB in November 2019 to cooperate in support of rural vitalization in the PRC, can potentially provide the DCG with a cloud computing platform, such as its Alibaba Cloud solution, and technical support. ADB. 2019. ADB, Alibaba Establish Strategic Partnership for Rural Vitalization in PRC. News release. 20 November.

Box 3: Key Features of the Proposed Asset Management System

Digital asset inventory and database on geographic information systems. Geographic information systems will be used for asset operations, with capabilities that include (i) service or asset location; (ii) records of asset alterations, performance monitoring and evaluation, maintenance activities, and expenses; (iii) formulation of a maintenance plan; and (iv) development of operating procedures. The inventory allows asset registration of new or existing infrastructure. These inputs, including total expenditure and nature of works, are provided by individual line agencies on a regular (real-time) basis as new infrastructure is added or when operation or maintenance works are completed. The database will be readily accessible by a decision support system. Annual audits of assets need to be performed to maintain the functional quality of the asset management system.

Interactive service delivery and automatic water leakage detection. Cloud computing has become popular for delivery services on a real-time basis. Cloud-based geographic information systems can be used to provide a real-time data model to realize environmental data management and establish an interactive water service delivery system, whereby users can report issues and problems to the service provider in real time. On the other hand, the service provider can readily locate the problem and contact the customer to fix the problem. The platform can also be used to collect revenue or pay water bills.

The water infrastructure installed by the project (such as the two new rural water supply plants in Sangzhuang and Jiulong) already has automatic leak-detection technology (acoustic detection method) embedded. Leaks will be identified by analyzing data obtained from detection devices installed along pipelines. If a water leak is detected in any section, the system initiates possible actions in response, such as automatically bypassing that network link or cutting water supply to that particular location.

Decision support system for asset acquisition and replacement. The decision support system is a programmed platform that can analyze the history of operation and maintenance of any particular asset and its associated costs over time. The system is linked to a cost–benefit analysis model, with additional inputs for present and projected demand. Recommendations made by the system can be used by the Dengzhou City Government to decide whether to continue with operation and maintenance of that particular asset, or replace it within an advance budget plan. However, this requires consultation with multiple stakeholders, including line agencies and beneficiaries. Consequently, the Dengzhou City Government needs to implement appropriate policies and strategies to support the decision-making process, including protocols to cover asset acquisition, operation and maintenance, overhaul, and disposal.

Guidelines and training for asset criticality and risk assessment. Asset management is a systematic and dynamic process. Support provided by the project will help gradually improve the quality of asset management and the decision support system. Operational guidelines for asset management will be prepared. As asset management is a collective responsibility that needs a participatory approach, capacity needs to be developed, mainly within line agencies.

Source: Asian Development Bank. People's Republic of China: Henan Dengzhou Integrated River Restoration and Ecological Protection Project.

135.   In addition to the water resources modeling, a climate change analysis was carried out to assess the climate risks involved. This was done by statistical downscaling of results from six Coupled Model Intercomparison Project Phase 5 protocol general circulation models,[48] specifically predictions of climate change from 2021 to 2050 under the representative concentration pathway (RCP) 4.5 and RCP8.5 climate change scenarios.[49] By 2050, it is projected that annual mean temperature in Dengzhou City will increase

---

48   World Climate Research Programme. CMIP Phase 5 (CMIP5).

49   CoastAdapt. What Are the RCPs? *Coastal Climate Change Infographic Series.* No. 1. National Climate Change Adaptation Research Facility. Queensland, Australia.

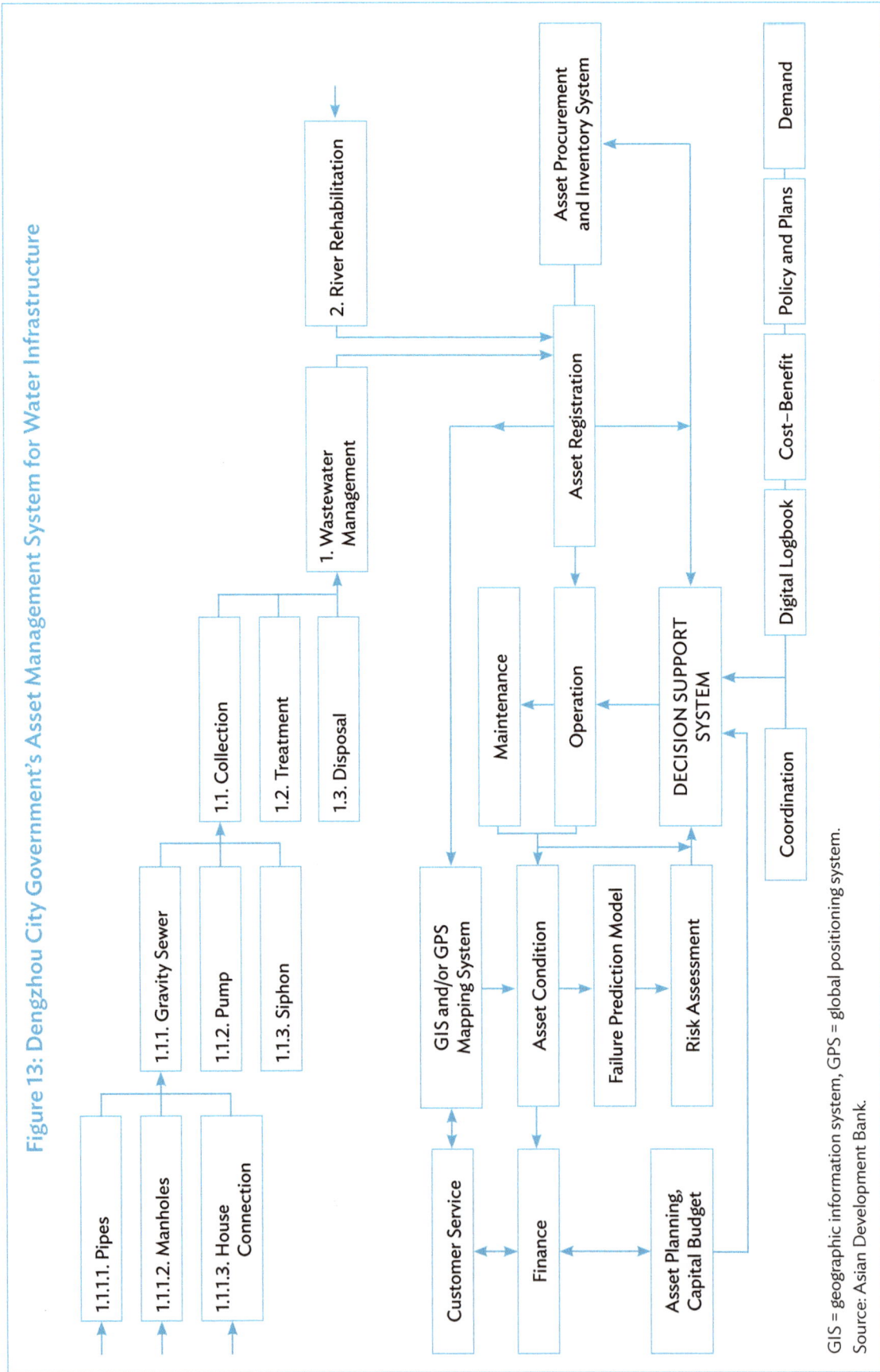

Figure 13: Dengzhou City Government's Asset Management System for Water Infrastructure

GIS = geographic information system, GPS = global positioning system.
Source: Asian Development Bank.

by 1.28°C under RCP4.5, or by 1.51°C under RCP8.5. The RCP4.5 scenario predicts annual precipitation will increase by 27.1 millimeters per year, or by 31.0 millimeters per year under the RCP8.5 scenario. The analysis also showed increased frequency and intensity of extreme rainfall events, which are likely to cause more severe storms and floods. The probability of drought events will increase, caused by higher temperatures and rates of evaporation.

2.   Pollution Reduction

136.   The estimated reduction in pollutant loading attributable to key project components is shown in Table 2. Open green space, pervious pavements, and wetlands in the Tuanbei New District are expected to achieve 55%–80% reduction of NPS pollution to the Tuan River. The project components will reduce pollutant loadings from the Tuan River Basin at the downstream Jituan station by 28% for NH_3-N, and by 19% for total phosphorus.

Table 2: Pollution Reduction by Key Project Components
(ton per year)

Project Component	Pollutant Load Reduction		
	NH_3-N	SS	TP
Proposed parks and associated wetlands		174	0.3
Tuanbei wastewater treatment plant	279		29.6
Township wastewater treatment plants	28		3.2
River restoration of the lower Tuan River		2,236	3.6
Total Reduction	**307**	**2,409**	**36.6**

NH_3-N = ammonia nitrogen, SS = suspended sediment, TP = total phosphorus.
Note: Numbers may not sum precisely because of rounding.
Source: Asian Development Bank. People's Republic of China: Henan Dengzhou Integrated River Restoration and Ecological Protection Project.

137.   Wastewater treatment facilities remove pollutants from industrial and municipal wastewater and reduce their impact on the environment. In the project, tertiary wastewater treatment utilizes technologies that remove nutrients (such as phosphorus and nitrogen) and improve the efficiency of wastewater treatment. These treatment facilities are designed to improve the water quality of runoff and effluents, and reduce pollutants entering receiving water bodies by 80%–95%.

138.   Dredging of the polluted sediment layer along 13.8 kilometers (km) of river with a sediment volume of 0.21 million cubic meters (mcm) will remove a significant proportion of pollutant loads accumulated on the bed of the Tuan River. The estimated quantities removed from the sediment of the Tuan River through the dredging process will be 350 tons of total phosphorus and 770 tons of total nitrogen.

139.   **Bio-shield interception of nonpoint source pollution.** The green buffer belts along the Tuan River will reduce the volume of direct runoff from farmlands to the river, trap sediments through porous media, and break down contaminants through biochemical activity. The estimated interception of NPS pollutants by the bio-shield buffer zone along the lower Tuan River under the project is shown in Table 3.

140.   In addition, the green buffer belts will improve air quality of the Tuan River Basin. It is estimated that the green buffer belts will absorb 55,482 kilograms (kg) of carbon dioxide per day and 3,312 kg of sulfur dioxide per month, and reduce airborne dust by 16.56 million–49.68 million kg per year.

Table 3: Pollutant Interception by Bio-shield Buffer Zone, Tuan River

Project	Length (km)	Size ('000 m^2)	Control Area of the Sub-catchment (ha)	Bio-shield NPS Pollutant Reduction (ton per year)			
				COD	TN	SS	TP
Restoration for the downstream part of the Tuan River	13.8	1,250	2,808	1,341	54	2,236	3.6

COD = chemical oxygen demand, ha = hectare, km = kilometer, m^2 = square meter, NPS = nonpoint source, SS = suspended sediment, TN = total nitrogen, TP = total phosphorus.
Source: Asian Development Bank. People's Republic of China: Henan Dengzhou Integrated River Restoration and Ecological Protection Project.

141.   Overall, the project's integrated urban–rural water management, with interventions in both urban and rural areas, will have a substantial effect in reducing pollutant loads and improving water quality in the lower Tuan River, as demonstrated by the estimates in Table 4.

Table 4: Summary of Pollutant Reduction by the Project

Item	COD	NH$_3$-N	TP	TN
Current pollution load in the Tuan River Basin (ton per year)	9,652	1,781	152	1,971
Reduction by project interventions (ton per year)	4,721	534	36	408
Reduction (%)	48	30	25	23

COD = chemical oxygen demand, NH$_3$-N = ammonia nitrogen, TN = total nitrogen, TP = total phosphorus.
Source: Asian Development Bank. People's Republic of China: Henan Dengzhou Integrated River Restoration and Ecological Protection Project.

3.   River Water Flow

142.   The flow in the Tuan River is directly related to the amount of water produced and the consumptive and non-consumptive uses in the watershed—i.e., the total amount of water moving off the watershed into the river. Each project intervention is aimed at conserving and saving water so that the minimum environmental flow with acceptable water quality is always maintained in the Tuan River. There are also several measures that are not part of the project but complement the project's objective such as limiting groundwater withdrawal for irrigation and water supply, and promoting river–groundwater recharge process. Other project impacts, including socioeconomic impacts, are not discussed in this report as the focus is the restoration of the Tuan River's health.

V

Conclusion and Way Forward

A. Summary of Urban–Rural Water Linkages for Dengzhou City

143.   Dengzhou City is a primarily agricultural region in the Tuan River Basin in Henan Province of the People's Republic of China (PRC). Rapid urbanization, combined with increasing point and nonpoint sources (NPSs) of pollution, has led to deteriorating environmental conditions along the Tuan River. Economic exploitation of natural resources and urbanization have (i) altered river basin hydrology; (ii) caused habitat loss and deforestation; and (iii) led to overextraction of groundwater, substandard water supply for rural areas, NPS pollution, overburdened solid waste disposal sites, aggravated flood hazards, and degraded water environments in both the urban and rural parts of the Tuan River Basin.

144.   Many urban–rural water linkages (URWLs) and dependencies were identified relating to the water resources and their poor environmental condition. Past efforts to better manage the water resources had limited success because they addressed small parts in isolation where particular problems were manifest, without giving due consideration to spatial and functional linkages, including water linkages between urban and rural areas. These are key environmental linkages that determine the ecological and environmental conditions of the Tuan River, and they need to be addressed by more integrated interventions. The three functional aspects of the key URWLs are land use, water withdrawal and water use, and pollution (water quality).

145.   Project interventions were proposed and implemented that address the problems in the urban and rural parts of Dengzhou City in a more integrated manner. Because of the linkages, the integrated interventions described in Chapter IV will greatly improve environmental conditions of the river and groundwater systems, and have direct and indirect beneficial impacts in urban and rural areas.

B. Features of the Project Interventions and Key Values Added

146.   The Asian Development Bank (ADB) is supporting the Dengzhou City Government (DCG) by means of a loan financing multiple measures that form a part of an extensive program as a demonstration to improve the serious water issues (footnote 3). The ADB components are consistent with integrated water resources management (IWRM) and, more specifically, address the URWLs. The project interventions will enhance water security, improve water quality and environmental conditions, promote sustainable and IWRM in the Tuan River Basin, and improve the livelihoods of urban and rural residents of Dengzhou City. Additional measures can then be selected to further beautify the urban river corridor to create an appealing environment for urban residents.

147.   Other key benefits of the project include (i) improving provision of public services in rural areas, (ii) improving environmental and sanitation conditions of rural areas, (iii) building an ecologically friendly rural living environment, and (iv) strengthening the capacity of the DCG. By promoting the URWLs approach,

the project contributes to improving rural livelihoods and welfare of rural residents, and restoring ecosystems and environmental conditions in urban and rural areas, including rehabilitation of the river system, the urban waterfront, and riparian zones. The project demonstrates IWRM best practices in the Yangtze River Economic Belt (YREB), identifying and improving URWLs, and supporting the DCG to manage water resources issues in a more holistic and sustainable way.

148.   The project also contributes the following values related to innovation and knowledge dissemination:

(i)   **Initiation of the rural vitalization plan.** The National Strategic Plan for Rural Vitalization, 2018–2022, issued by the PRC's State Council in September 2018, was intended to be implemented by local governments with the overall objectives of supporting thriving enterprises in rural areas, providing more pleasant living environments, facilitating greater social inclusiveness, and fostering more effective governance (footnote 37). The ADB project serves as an innovative and replicable demonstration pilot project for the PRC's rural vitalization plan, adding value by introducing and applying concepts of urban–rural integration and emphasizing the importance of URWLs in restoring river health. The project is part of a larger investment plan in the water and environment sectors by the DCG; thus, the demonstration of the concepts and their application in the project will have wider implications as the DCG implements its investment program.

(ii)   **Contribution to the Yangtze River Economic Belt.** Located within the Yangtze River Basin, the project in Dengzhou City is a significant contribution by ADB in its commitment to supporting the PRC to achieve the principal goals of its YREB Development Plan, 2016–2030 (footnote 43). It also demonstrates the potential of the URWLs approach for river rehabilitation and related social welfare for other river systems, not just those included in the YREB.

(iii)   **Integrated water resources management and climate change.** The project introduces innovative planning, design, and implementation methods based on IWRM to improve water quality, environmental conditions, and social welfare and equity in the Tuan River Basin. It advocates river and water resources management approaches that recognize spatial water linkages and the need for urban and rural integration of corrective measures. It adopts a cross-sector approach to water resources and environmental management, (a) with integration of various interventions in river basin management, water supply, wastewater treatment, solid waste management (SWM), wetland development, afforestation, bio-shield belts and other green measures, and ecological river rehabilitation; and (b) with provision for monitoring and evaluation. Effects of climate change induced by global warming were considered in the design of project interventions.

(iv)   **Low-impact development and/or sponge city approach.** The project adopts the low-impact development (LID) or sponge city approach in the design of open green space, urban infrastructure, and use of wetlands. The LID design of project interventions will help attenuate stormwater runoff to reduce flood risk; reduce amounts of pollutant-laden stormwater draining into local receiving waters, thereby improving water quality downstream; increase infiltration of rainfall and stormwater runoff, thereby enhancing groundwater recharge; restore aquatic habitat; and provide more attractive neighborhood landscapes, with opportunities for community recreation and relaxation.

(v)   **Use of bio-shields.** The project includes the creation of riparian vegetated buffer belts (bio-shields), with trees, shrubs, and grasses to intercept pollutants from on-farm NPSs draining to the Tuan River, and stabilize the riverbanks. The buffer zone plays beneficial roles in biodiversity conservation, landscape improvement, and soil erosion control.

(vi)   **Solid waste management.** The project includes a trial SWM subproject in a rural town to demonstrate good practice in SWM using advanced technology that reduces the volume of solid waste at the source and, therefore, lessens the risk of landfill leachate contaminating the soil and the groundwater.

(vii)   **River rehabilitation.** The project includes dredging of accumulated contaminated sediments deposited in the riverbed, and excavation to remove old earthen bunds and discarded construction materials that obstruct the natural river flow. The project implements eco-friendly riverbank protection and flood protection embankments. These initiatives will remove polluted sediment deposits to improve water quality and enhance the river flood conveyance capacity of the lower Tuan River.

C.   Way Forward and Upscaling Experience Gained

149.   The essence of urban–rural integration is to ensure that the benefits of development and growth reach all citizens—whether they move to megacities, live in smaller towns and cities, or in rural areas. Improved services, infrastructure, and living conditions will empower rural areas to contribute more to regional economic development. Local and regional authorities should implement smart solutions that better connect urban and rural areas through more integrated governance and land use planning, and facilitate greater access to modern technology to boost production and enterprise, especially in rural areas. A better balance between service provision in urban and rural areas would arise from more unified administration of services, which can be enabled by innovative use of modern technology.

150.   To enable rural areas to prosper, an integrated social development approach for urban and rural areas is needed. For this, governments have to take a leading planning role, providing basic facilities, amenities, and infrastructure. Governments should also assume the role of facilitator, engaging the private sector as an effective development partner, and should stimulate diversification of the rural economic base by promoting resource-based industries and other economic activities based upon the latent strengths of these areas.

151.   The IWRM approach is a prerequisite for the URWLs approach, with the river basin as the preferred spatial scale of planning, development, and management. IWRM follows a bottom-up approach and encourages community participation in all processes. The holistic view of IWRM jointly considers quantity and quality of both surface water and groundwater. The project described in this document is a case study of this integrated approach.

152.   An important element in this project is the protection and restoration of the rural ecology. Natural resources must be protected, and their economic exploitation managed sustainably. In many rural areas, this will mean the introduction of more eco-friendly agricultural practices and cleaner production methods. The use of fertilizer and pesticides in rural areas should be minimized, and the collection, transportation, or reuse of organic wastes should be promoted as an alternative. Recycling and reuse of agriculture and forestry by-products and livestock manure should be evaluated. This all requires strengthening of relevant institutions, and making them more flexible to deal with cross-sector strategies. The basic regulations for this are in place in the PRC. As a related matter, the new river chief system should be expanded and extended more at the village level.

153.   **The 14th Five-Year Plan, 2021–2025 and beyond.** The 5-year development plans in the PRC have continuously evolved to address emerging challenges in water and environmental management. The PRC's 13th Five-Year Plan, 2016–2020 proposes a cleaner and greener economy, with stronger commitments to environmental management and protection, clean energy and emissions controls, ecological protection and security, and the development of green industries.[50] It provides a strong and clear vision on environmental protection by learning from the implementation of the 12th Five-Year Plan, 2011–2015. It prioritizes the establishment of unit-based management of water quality; holistic strategies to tackle groundwater pollution; basin-wide strategies; protection of surface water bodies; and improvement of water quality in urban rivers and lakes, estuaries, and coastal areas. The 13th Five-Year Plan sets measurable indicators to evaluate the goals, such as the target that more than 70% of the PRC's surface water reaches class III water quality or the equivalent by 2020. The 14th Five-Year Plan, 2021–2025 is currently under formulation and will be crucial in shaping the

---

50   Government of the PRC. 2016. *13th Five-Year Plan for Economic and Social Development of the People's Republic of China (2016–2020).* Beijing.

future for the PRC. The country has risen from low-income to upper middle-income status, and the PRC will be more sensitive toward meeting environmental goals that support its objective of becoming a prosperous nation in the coming years.

154.   One of the objectives of the URWLs approach in the Henan project was to use the experience gained as an example of how to address these linkages in other parts of the PRC. The Dengzhou project is located within the Yangtze River Basin, which is relatively blessed with abundant water resources. Applying the URWLs approach in a water-scarce river basin will be a greater challenge. One such river basin might be the Yellow River Basin, in the much drier northern parts of the PRC. The high erodibility of soil on the Loess Plateau at the heart of the Yellow River Basin presents special challenges. ADB, in cooperation with the Government of the PRC, is developing a project related to the Yellow River ecological corridor program.

155.   **Yellow River Basin.** The Yellow River flows through nine provinces and autonomous regions—(from west to east) Qinghai, Sichuan, Gansu, Ningxia Hui Autonomous Region, Inner Mongolia Autonomous Region, Shaanxi, Shanxi, Henan, and Shandong—before discharging to the Bohai (Yellow) Sea. The 5,500 km Yellow River feeds about 12% of the PRC's population; irrigates about 15% of its arable land; generates 14% of national gross domestic product; and supplies water to more than 60 cities, including Beijing and Qingdao municipalities located outside the river basin. This is the second largest river basin in the PRC after the Yangtze River Basin; of all the world's rivers, the Yangtze River has among the highest suspended sediment concentrations and sediment transport.

156.   Because of sediment deposition in the lower reaches of the Yellow River, the riverbed continuously aggrades and has caused frequent severe floods and changes in the river course. The flood in 1887 was the worst in recent centuries, killing more than 900,000 people. But more recent flood events have occurred that have caused substantial damage despite great expenditure on flood control works. A main source of the sediment, and a main cause of flooding, is the poor agricultural practices in the Loess Plateau.

157.   The Yellow River Conservancy Commission has played a key role in managing water resources in the river basin, but it lacks the mandate to coordinate the overall development in multiple sectors within the river basin. The experience with the URWLs project in Henan Province can potentially support the Yellow River Conservancy Commission and local governments in better addressing the specific management and conservation issues of the Yellow River Basin.

158.   **Other potential implications.** ADB's continuing engagement with the Government of the PRC in meeting environmental challenges could be strengthened by considering the URWLs approach, particularly under the YREB and rural vitalization programs, and any potential future initiatives including the Yellow River Basin program. That would help further develop the approach and refine its application in the PRC and in other ADB developing member countries (Box 4). As the PRC's long-term development partner, ADB can contribute to achieving and enriching the country's environmental goals, particularly during the 14th Five-Year Plan period, by sharing knowledge and ideas derived from ADB's broad experience of supporting sustainable development and environmental conservation.

159.   As the world grapples with the disruption caused by the coronavirus disease (COVID-19) global pandemic, the balance between human society and the natural environment, routinely taken for granted, has been exposed as fragile. This report, directed toward restoring river health in degraded environments, is highly relevant to the objective of redressing this balance. The COVID-19 pandemic has further emphasized that human society must work with nature, not against it, in order to achieve the development goals and secure the benefits that countries seek to achieve.

Box 4: Potential Application of the Urban–Rural Water Linkages Approach in the Pacific Islands

The urban–rural water linkages approach can help address environmental degradation issues, as well as pilot integrated solutions, in the Pacific islands. Small island developing states are at the nexus of ocean health, climate change, urban development, natural disasters, service delivery gaps, and inequality along the urban–rural continuum. Rainfall is highly variable in many islands, and most of the population still rely on limited surface water or groundwater resources (confined between lagoons and seas), which are highly vulnerable because of their natural settings and uncoordinated development.

In Fiji, for example, the United Nations Economic and Social Commission for Asia and the Pacific estimated that annual losses arising from extreme weather events could reach 6.5% of the country's gross domestic product by 2050.[a] These challenges are exacerbated by inadequate planning capacity, lack of inclusiveness, and continuing destruction of the coastal and marine ecosystems through unsustainable land use and coastal development, overfishing, solid waste pollution, and climate change-induced natural disasters.

Through its Pacific Approach, 2016–2020, the Asian Development Bank has been supporting island states in implementing innovative solutions to strengthen disaster preparedness, reduce vulnerabilities, and build resilience.[b] Tangible examples to address these issues more holistically include ongoing interventions in Fiji and Kiribati to improve integrated water supply and sanitation services. Investment programs in these countries look more closely at alternative sources of water, including desalination, and decentralized sanitation and hygiene practices to reduce waterborne diseases. Tapping into existing opportunities of linking ocean cities with their landscape and seascape ecosystems, the urban–rural water linkages approach can be a good example for promoting social and economic benefits within the communities at large.

[a] United Nations Economic and Social Commission for Asia and the Pacific (ESCAP). 2018. Re-Naturing Urbanization. *Ocean Cities of the Pacific Islands Policy Brief.* No. 3. Bangkok: ESCAP Environment and Development Division / Suva, Fiji: ESCAP Subregional Office for the Pacific.
[b] Asian Development Bank (ADB). 2018. *Country Operations Business Plan: 11 Small Pacific Island Countries, 2019–2021.* Manila; and ADB. 2016. *Pacific Approach, 2016–2020.* Manila.

Source: ADB.

www.ingramcontent.com/pod-product-compliance
Lightning Source LLC
Chambersburg PA
CBHW051658210326
41518CB00026B/2623